Intelligent Systems, Control and Automation: Science and Engineering

Volume 64

Series Editor

S. G. Tzafestas

For further volumes:
http://www.springer.com/series/6259

Matjaž Mihelj · Janez Podobnik

Haptics for Virtual Reality and Teleoperation

Springer

Matjaž Mihelj
Faculty of Electrical Engineering
University of Ljubljana
Ljubljana
Slovenia

Janez Podobnik
Faculty of Electrical Engineering
University of Ljubljana
Ljubljana
Slovenia

ISBN 978-94-017-8447-4 ISBN 978-94-007-5718-9 (eBook)
DOI 10.1007/978-94-007-5718-9
Springer Dordrecht Heidelberg New York London

Printed on acid-free paper

Springer is part of Springer Science+Business Media (www.springer.com)

Contents

Chapter 1
Introduction to Virtual Reality

Virtual reality is composed of an interactive computer simulation, which senses the user's state and operation and replaces or augments sensory feedback information to one or more senses in a way that the user gets a sense of being immersed in the simulation (virtual environment) [1]. Thus, it is possible to identify four key elements of virtual reality: virtual environment, sensory feedback (in response to user activity), interactivity and immersion.

A computer generated *virtual environment* presents descriptions of objects within the simulation and the rules as well as relationships that govern these objects. Viewing of the virtual environment through the system, which displays objects and enables interaction resulting in immersion, leads to virtual reality.

Sensory feedback is a necessary element of virtual reality. The virtual reality system provides a direct sensory feedback to users based on their pose (position and orientation) and actions. In most cases vision is the most important sense through which the user perceives the environment. In order for the sensory feedback information to correspond to the current user pose, it is necessary to track user movement. Pose tracking defines computer-based measurement of the position and orientation of an object in the physical environment.

For virtual reality to become realistic, it must be responsive to user actions; it has to be *interactive*. The ability to influence the unfolding of events in a computer-generated environment is one form of interaction. Another is the ability to modify the perspective within the environment. A *multiuser environment* represents an extension of interactive operation and allows more users to interactively share the same virtual space and simulation. A multiuser environment must allow interaction between users.

When a user operates in the same environment as other users, it is important to perceive their presence in this. The notion of an *avatar* describes the representation of the user in a virtual environment. The avatar is a virtual object that represents the user or a physical object within the virtual environment.

Immersion can be roughly divided into physical (sensory) and mental. Immersion represents a sense of presence in an environment. Physical immersion is the basic characteristic of virtual reality and represents a physical entry into the system.

M. Mihelj and J. Podobnik, *Haptics for Virtual Reality and Teleoperation*, Intelligent Systems, Control and Automation: Science and Engineering 64,
DOI: 10.1007/978-94-007-5718-9_1,

A virtual reality system must be able to establish at least minimal physical immersion. Synthetic stimuli that stimulate one or more of the user's senses in response to the pose (position and orientation) and actions of the user are generated by different display devices; this does not mean that they must cover all senses and the entire body. In general the virtual reality system creates images of a virtual environment in visual, aural and haptic forms. Visual, auditory and haptic information needs to adapt to changes in the scene according to the user's movement.

Mental immersion represents involvement in the virtual environment (engagement, expectation). A definition of mental immersion requires the user to be so occupied with his existence within the virtual space as to stop questioning whether it is real or fake. The level of mental immersion is affected by factors such as a virtual reality scenario, the quality of displays and rendering and the number of senses being stimulated by the virtual reality display system. Another major factor affecting immersion is the delay between the user's actions and the responses of the virtual reality system. If the delay is too long (*being too long* depends on the display type—visual, aural or haptic), it can destroy the effect of mental immersion. The level of the desired mental immersion changes with the purpose of the virtual reality. If the virtual reality experience is intended for entertainment, a high level of mental immersion is desired. However, a high degree of mental immersion is often not necessary, possible or even desirable.

Synthetic stimuli usually occlude stimuli originating from the real environment. This reduces the mental immersion in the real environment. The degree to which real stimuli are replaced by synthetic ones and the number of senses which are *fooled* by synthetic stimuli affect the level of physical immersion in the virtual environment. In turn, this affects the level of mental immersion.

Virtual reality is strongly related to other concepts such as augmented reality and telepresence.

Augmented reality represents an extension of virtual reality with superposition of synthetic stimuli (computer generated visual, aural or haptic stimuli) over stimuli that originate from real objects in the environment that directly or indirectly (via display) interact with the user. In this regard augmented reality is more general than virtual reality, since it integrates virtual reality with real images. Augmented reality usually allows the user to perceive otherwise unperceivable information (for example, a synthetic view of information originating from inside the human body superimposed over the appropriate point on the body surface).

Telepresence represents the use of virtual reality system to virtually move the user to another location. Telepresence represents the ability of interaction with a real and remote environment from the user's perspective. There are no limitations regarding the location of the remote environment. While telepresence refers to existence or interaction that includes a remote connotation, *teleoperation* in general indicates operation of a machine (often a robot) at a distance.

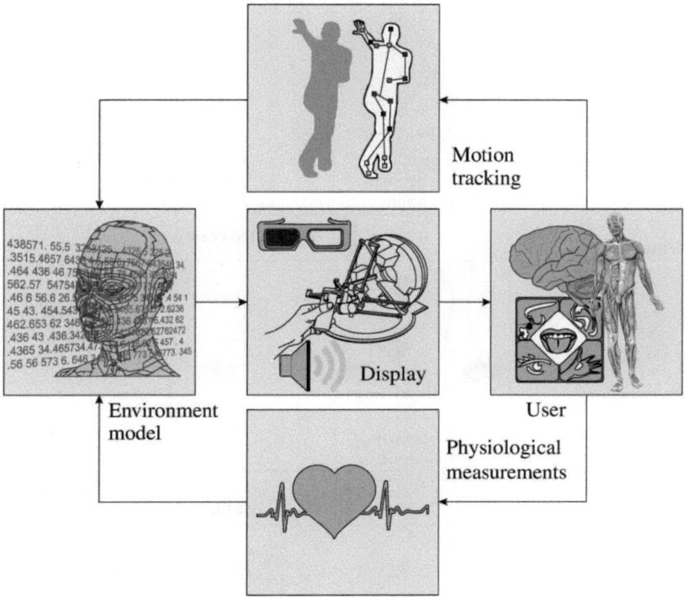

Fig. 1.1 A feedback loop is one of the key elements of a virtual reality system. The system has to be responsive to user actions. In order to increase user involvement, the user's psychological state can be assessed and taken into consideration when adapting the virtual environment

1.1 Virtual Reality System

Virtual reality relies on the use of a feedback loop. Figure 1.1 shows the feedback loop, which allows interaction with the virtual reality system through user's physical actions and detection of user's psychophysiological state. In a *fast* feedback loop the user directly interacts with the virtual reality system through motion. In a *slow* feedback loop, the psychophysiological state of the user can be assessed through measurements and analysis of physiological signals and the virtual environment can be adapted to engage and motivate the user.

The virtual reality system enables exchange of information with the virtual environment. Information is exchanged through the interface to the virtual world. The *user interface* is the gateway between the user and the virtual environment. The user interface defines how the user interacts with the virtual environment and how the virtual environment manifests itself to the user. Ideally, the gateway would allow transparent communication and transfer of information between the user and the virtual environment.

Figure 1.2 shows the flow of information within a typical virtual reality system. The virtual environment is mapped to a representation which is then rendered and displayed to the user through various displays. The rendering process selects the perspective based on the movement of the user, allowing immersion in the virtual

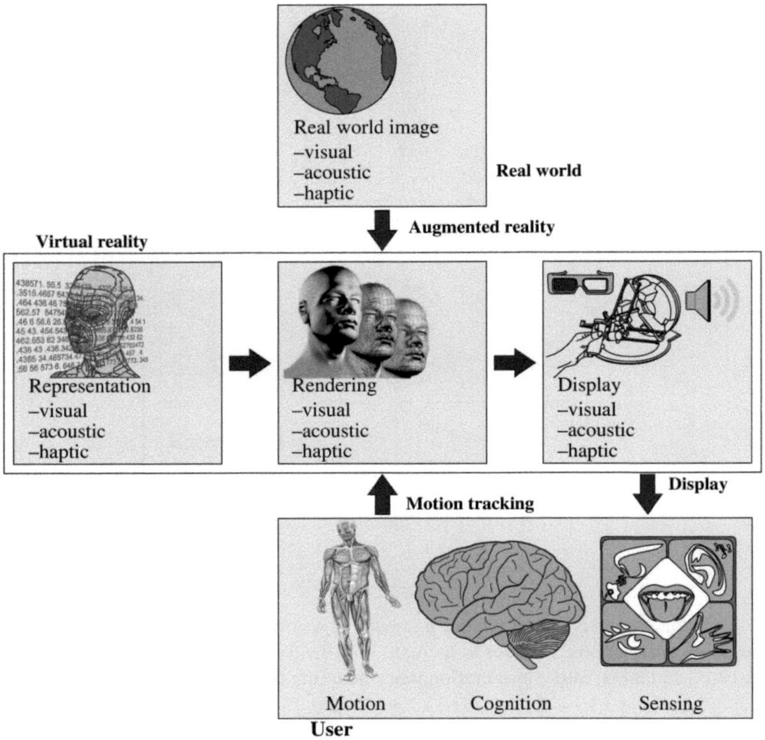

Fig. 1.2 Flow of information within a typical virtual reality system

environment. In an augmented reality system the display of the virtual environment is superimposed over the image of the real environment. The user can interact with and affect the virtual environment through the user interface.

1.1.1 Virtual Environment

Virtual environment is determined by its content (objects and characters). This content is being displayed through various modalities, visual, aural and haptic and perceived by the user through vision, hearing and touch.

Just like objects in the real world, also objects in a virtual environment have their properties such as shape, weight, color, texture, density and temperature. These properties can be observed using different senses. The color of an object, for example, is perceived only in the visual domain, while its texture can be perceived both in visual as well as haptic domains.

The content of the virtual environment can be grouped into categories. *Environment topology* describes the surface shape, areas and features. Actions in a virtual environment are usually limited to a small area within which the user can move. *Objects* are three-dimensional forms which occupy space in the virtual environment. They are entities that the user can observe and manipulate. *Intermediaries* are forms, which are controlled via interfaces, or avatars of users themselves. *User interface elements* represent parts of the interface that resides within the virtual environment. These include elements of virtual control such as virtual buttons, switches or sliders.

A model of an object in a virtual environment must include a description of its dynamic behavior. This description also defines the object's physical interaction with other objects in the environment. Object dynamics can be described based on various assumptions, which then determine the level of realism and the computational complexity of simulation.

Static environment, for example, consists of stationary objects, around which the user moves. Real-world physical laws are not implemented in the virtual environment. The computational complexity in this case is the lowest.

On the other hand, *Newtonian physics* represents an excellent approximation of real-world physics and includes conservation of momentum as well as action and reaction forces. Objects behave realistically, but computational complexity increases significantly. This can be simplified by a *set of rules* that are less accurate than Newtonian physics, but often describe developments in a way that seems natural to most humans.

Newtonian physics can be upgraded with physical laws that describe events in an environment that is beyond our perceptions. These laws apply either to micro (molecules, atoms) or macro environments (galaxies, universe) and are defined by quantum and relativistic physics.

1.2 Human Factors

Humans perceive their environment through multiple distinct sensory channels. These enable perception of electromagnetic (vision), chemical (taste, smell), mechanical (hearing, touch, vestibular sense) and heat stimuli. Most of these stimuli can be reproduced artificially using the virtual reality system, though chemical stimuli are rarely implemented.

All stimuli, natural or artificial, are finally filtered through the human sensory system. Therefore, the virtual reality system and the virtual environment must take into account the characteristics of sensing, which are physiological, psychological and emotional in nature. An important aspect of human cognition is also the ability to generalize, which allows grouping of objects and ideas with similar characteristics.

In order to create convincing experience, at least a basic understanding of the physiology of human sensing is required. Since a detailed analysis is beyond the scope of this book, only visual, auditory, tactile, kinesthetic and vestibular senses will be briefly presented.

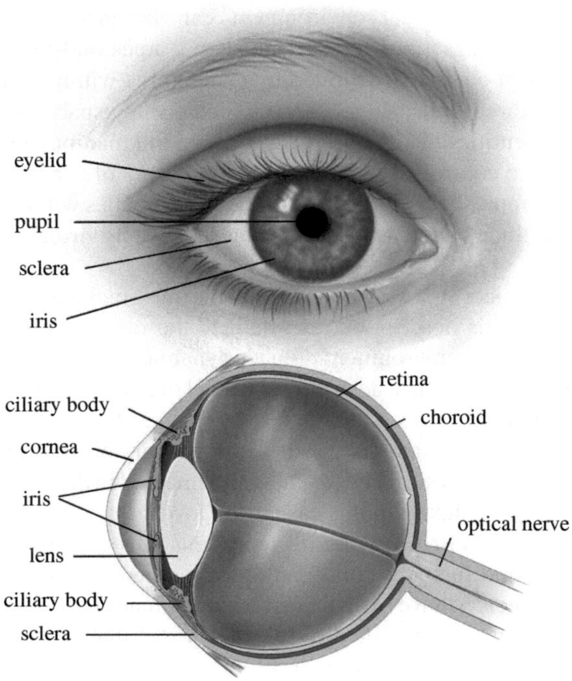

Fig. 1.3 Human eye anatomy

1.2.1 Visual Perception

Visual perception is the ability to interpret information from visible light reaching the eye. The various physiological components involved in vision constitute the visual system (Fig. 1.3). This system allows individuals to assimilate information from the environment. This information includes object properties such as color, texture, shape, size, position and motion.

Most visual displays are two-dimensional. They lack the third dimension, depth. Thus, while object color and texture can be displayed easily, presentation of other characteristics is limited to the visual plane of the display. Understanding of human depth perception is necessary to be able to trick the human visual system into seeing depth on a two-dimensional display. Depth can be inferred from different indicators called depth cues.

Monocular depth cues (Fig. 1.4) can be observed in a static view of the scene. *Occlusion* is the cue, which occurs when one object partially covers a second object. *Shading* gives a sense of object shape, while shadow indicates position dependence between two objects. By comparing the *size* of two objects of the same type, it is possible to determine their relative distance; absolute distance can be inferred from

Fig. 1.4 Monocular depth cues are important for depth assessment (linear perspective, shadows, occlusion, texture gradient, and horizon)

our previous experience. *Linear perspective* represents the observation of parallel lines converging in a single point and is relevant primarily for objects constructed from straight lines. *Surface texture* of more distant objects is less pronounced than the texture of closer objects, whereas the retina cannot separate details in the texture at larger distances.

Binocular depth cues rely on a pair of eyes. *Stereopsis* is the process leading to the sensation of depth from the two slightly different projections of the world onto the retinas of each eye. The differences in the two retinal images are called binocular disparity and arise from the different positions of the eyes in the head (Fig. 1.5). Stereopsis is important for manipulation of objects. *Convergence* is a binocular oculomotor cue for depth perception (Fig. 1.6). By virtue of stereopsis, the two eye balls focus on the same object. In doing so, they converge, stretching the extraocular muscles. Kinesthetic sensations from these extraocular muscles also help in depth perception. The angle of convergence is smaller, when the eye is fixating on far away objects.

Motion parallax derives from the parallax, which is generated by varying the relative position of the head and the object. Depth information comes from the fact that objects which are closer to the eye appear to move faster through the retina than the more distant ones. *Kinetic depth perception* is determined by dynamically changing object size. As objects in motion become smaller, they appear to recede into the distance or move farther away. Objects in motion that appear to be getting larger seem to be coming closer. Use of kinetic depth perception allows the brain to calculate time-to-contact distance at a particular velocity.

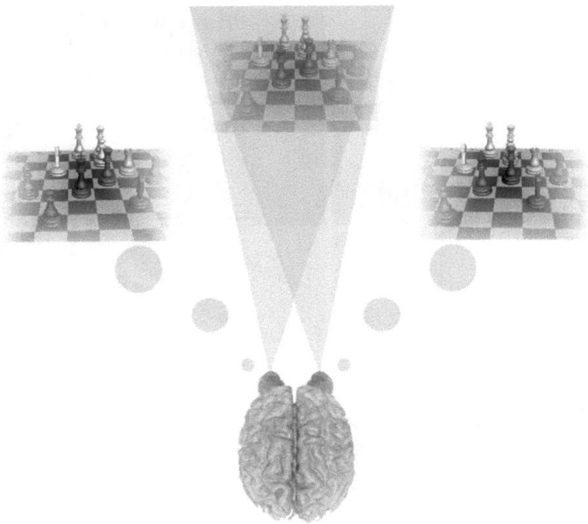

Fig. 1.5 The concept of stereopsis

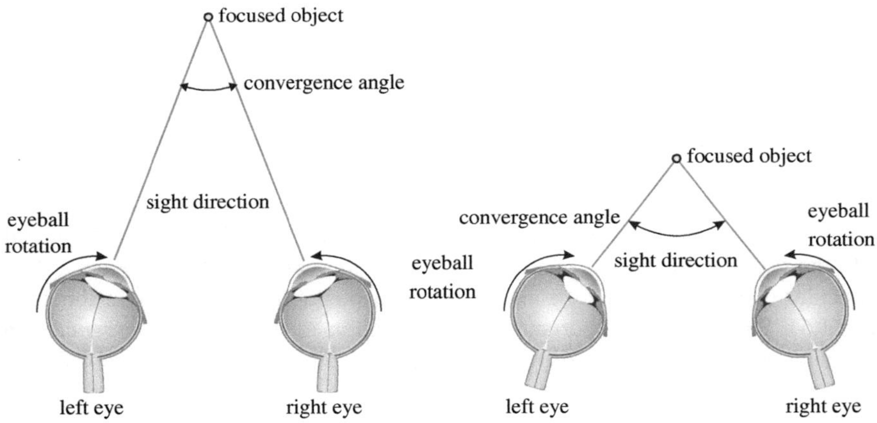

Fig. 1.6 Convergence and accommodation. Eyeballs rotate toward the object being observed (convergence). At the same time the lens is deformed in a way to allow focusing of the object (accommodation)

In case of conflicts between different depth cues, stereopsis prevails over other cues. Motion parallax is also a strong depth cue. Within monocular cues, occlusion is the strongest. Physiological cues (convergence) are the weakest. Any combination of depth cues can be used in order to create a convincing illusion of depth. Characteristics of visual perception are those that define properties of visual displays.

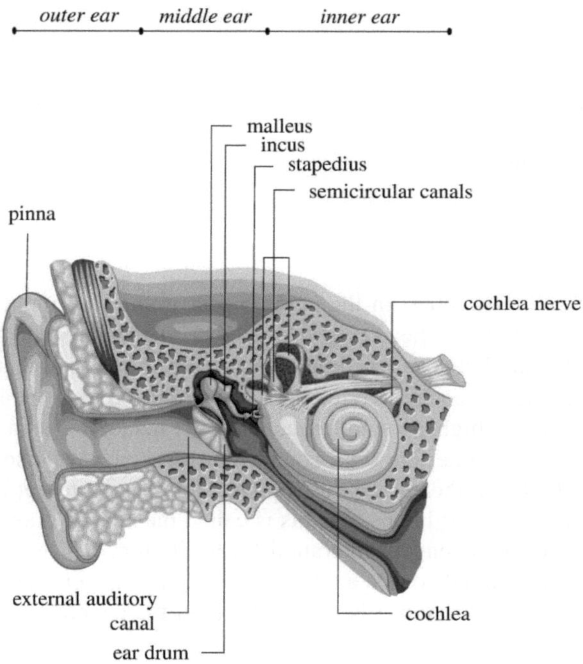

Fig. 1.7 Human ear anatomy

1.2.2 Aural Perception

Aural perception is the ability to interpret information from sound waves reaching the ear (Fig. 1.7). Since a virtual world is usually a three-dimensional environment, it is important to consider three-dimensional sound in virtual reality. *Sound localization* represents a psycho-acoustic phenomenon, which is the listener's ability to identify the location or origin of the detected sound (its distance and direction), or acoustical engineering methods for simulating the placement of an auditory cue in a virtual three-dimensional space. Sound localization cues are analogous to visual depth cues. General methods for sound localization are based on binaural and monaural cues.

Monaural localization mostly depends on the filtering effects of the human body structures. In advanced audio systems these external filters include filtering effects of the head, shoulders, torso and outer ear and can be summarized as a head-related transfer function. Sounds are filtered depending on the direction from where they reach various human body structures. The most significant filtering cue for biological sound localization is the *pinna notch*, a notch filtering effect resulting from interference of waves reflected from the outer ear. The frequency that is selectively notch filtered depends on the angle from which the sound strikes the outer ear.

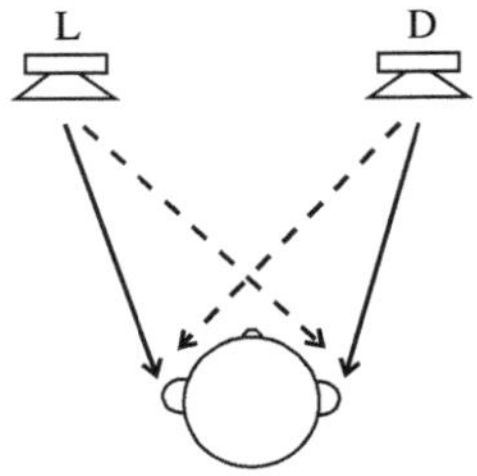

Fig. 1.8 Binaural localization relies on the comparison of auditory input from two ears, one on each side of the head. Interaural time and level differences aid in localization of the sound source azimuth

Binaural localization relies on the comparison of auditory input from the two ears, one on each side of the head (Fig. 1.8). The primary biological binaural cue is the split-second delay between the time when sound from a single source reaches the near ear and when it reaches the far ear. This is referred to as the interaural time difference. However, at higher sound frequencies, the size of the head becomes large enough that it starts to interfere with sound transmission. With the sound source on one side of the head, the ear on the opposite side begins to get occluded (thus receiving sounds at a lower intensity); this is called the head-shadowing effect and results in a frequency-dependent interaural level difference. These cues will only aid in localizing the sound source azimuth (the angle between the source and the sagittal plane), not its elevation (the angle between the source and the horizontal plane through both ears).

Distance cues do not rely solely on inter-aural time differences or monaural filtering. Distance can theoretically be approximated through interaural amplitude differences or by comparing the relative head-related filtering in each ear. The most direct distance cue is sound amplitude, which decays with increasing distance. However, this is not a reliable cue, since it is generally not known how strong the sound source is.

In general, humans can accurately judge the sound source azimuth, less accurately its elevation and even less accurately the distance. Source distance is qualitatively obvious to a human observer, when a sound is extremely close or when sound is echoed by large structures in the environment.

Sound localization describes the creation of the illusion that the sound originates from a specific location in space. Localization is based on different cues. One possibility is implementation of a head-related transfer function, which alters signal properties to make it seem, as if the sound is originating from a specific location.

In general, the human ability to localize sounds is relatively poorly developed. Therefore, it is necessary to use strong and unambiguous localization cues in a virtual reality system.

1.2.3 Haptic Perception

Haptic perception represents active exploration and the process of recognizing objects through touch. It relies on the forces experienced during touch. Haptic perception

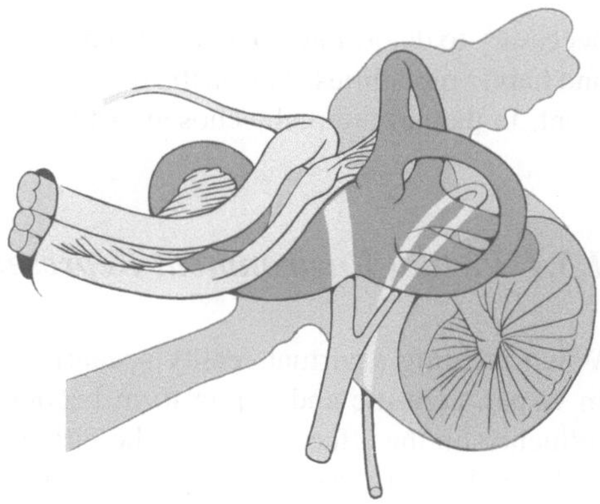

Fig. 1.9 Vestibular system, located in inner ear, contributes to human balance and sense of spatial orientation

involves a combination of somatosensory perception of patterns on the skin surface and kinesthetic perception of limb movement, position and force. People can rapidly and accurately identify three-dimensional objects by touch. They do so through the use of exploratory procedures, such as moving the fingers over the outer surface of the object or holding the entire object in the hand. The concept of haptic perception is related to the concept of extended physiological proprioception according to which, when using a tool, perceptual experience is transparently transferred to the end of the tool. Haptic perception is discussed in further detail in Chap. 3 *Human haptic system*.

1.2.4 Vestibular Perception

The vestibular system, which contributes to human balance and sense of spatial orientation, is the sensory system that provides the dominant input about the movement and equilibrioception. Together with the cochlea, a part of the auditory system, it constitutes the labyrinth of the inner ear (Fig. 1.9). As human movements consist of rotations and translations, the vestibular system comprises two components: the semicircular canal system, which indicates rotational movements; and the otoliths, which indicate linear acceleration. The vestibular system sends signals primarily to the neural structures that control eye movements and to the muscles that keep a body upright.

1.3 Virtual Environment Representation and Rendering

Rendering is the process of creating sensory images depicting the virtual environment. Images must be updated fast enough (real-time rendering) so that the user gets a sense of continuous flow. Creating sensory images consists of two steps. First, it is necessary to determine how the virtual environment should look in visual, acoustical and haptic modalities. This is the representational level of creating a virtual environment. In the next step, the chosen virtual environment representation is rendered.

1.3.1 Virtual Environment Representation

When creating a virtual reality scenario, methods of presentation of information in visual, acoustic and haptic form become important. They may have significant influence on the effectiveness of the virtual reality simulation.

Virtual reality usually strives for a realistic representation of the environment. The more realistic is the representation, the more likely it is that there will be no ambiguity in interpretation of information.

1.3.1.1 Visual Representation in Virtual Reality

Visual perception is of primary importance when gathering information about the environment and the appearance of nearby objects.

Vision in a virtual environment enables determination of the user's position relative to the *entities*—avatars and objects—in a virtual space. This is important for navigation through space as well as manipulation of objects and interaction with other users in this environment. In addition to perceiving the position of entities, it is possible to distinguish their shape, color and other characteristics, based on which they can be recognized or classified.

Vision is characterized as a remote sense, since it enables perception of objects that are beyond our immediate reach. It allows observation of things that are not in direct contact with the body. When they are perceived, it is immediately possible to estimate their pose and visual characteristics.

Vision, in addition to recognizing entities in the virtual environment, also allows recognition of gestures for communication purposes. In a multiuser environment, communication between users is possible using simple gestures simulated through avatars.

1.3.1.2 Auditory Representation in Virtual Reality

Sound increases the feeling of immersion in a virtual environment. Ambient sounds, which provide cues on the size and nature of the environment and mood in the room, and sounds associated with individual objects, form the basis for the user's understanding of space.

Sound attracts attention. At the same time, it also helps determine object position in relation to the user. Like vision, hearing is classified as a remote sense. However, unlike vision, sound is not limited by head orientation. The user perceives the same sound irrespective of the orientation of the head.

Temporal and spatial characteristics of sound are different from those of visual information. Although what we see exists in space as well as time, vision emphasizes the spatial component of the environment. In contrast, sound emphasizes in particular the temporal component. Since sound exists mainly in time, the timing of sound presentation is even more critical than the timing of image presentation.

Sound in virtual reality can be used to increase the sense of realism of the virtual environment, to provide additional information or to help create a mood. Realistic sounds help establish mental immersion, but can also provide practical information about the environment.

1.3.1.3 Haptic Representation in Virtual Reality

Different characteristics of the real environment are perceived through the haptic sense. The objective of using haptic displays is to represent the virtual environment as realistically as possible. Abstract haptic representations are rarely used, except in interactions with scaled environments (e.g. nanomanipulation), for sensory substitution and for the purpose of avoiding dangerous situations. In interactions with scaled environments, the virtual reality application may use forces perceivable to humans, for example, to present events unfolding at the molecular level.

Information that can be displayed through haptic displays includes object features such as texture, temperature, shape, viscosity, friction, deformation, inertia and weight. Restrictions imposed by haptic displays usually prevent the use of combinations of different types of haptic displays.

In conjunction with visual and acoustic presentations, the haptic presentation is the one that the human cognitive system most relies on in the event of conflicting information. Another important feature of haptic presentations is its local nature. Thus, it is necessary to haptically render only those objects that are in direct reach of the user. This applies only to haptic interactions, since visual and auditory sensations can be perceived at a distance, which is out of immediate reach.

Haptic interaction is frequently used for exploration of objects in close proximity. Force displays are used in virtual reality for displaying object form, for pushing and deforming objects. Simulation of virtual environment defines whether the applied force results in deflection or movement of objects. Haptic displays can be divided into three major groups.

Force displays are especially useful for interaction with a virtual environment, for the control or manipulation of objects and for precise operations.

Tactile displays are especially useful in cases where object details and surface texture are more important than overall form.

Passive haptic feedback can be based on the use of control props and platforms. These provide a passive form of haptic feedback.

1.3.1.4 Sensory Substitution

Due to technical limitations of the virtual reality system, the amount of sensory information transmitted to the user is often smaller than in a real environment. These limitations can partially be compensated with a sensory substitution resulting in a replacement of one type of display with another (for example, sound instead of a haptic display can be used to indicate contact). Some substitutions are used for similar sensations; it is, for example, possible to apply vibrators at the user's fingertips to provide information about the contact with an object. In general, sensory substitution is used when the technology that would allow presentation of information in its natural form is too expensive or nonexistent.

1.3.2 Virtual Environment Rendering

Rendering generates visual, audio and haptic images to be presented to the user through the corresponding displays. Hardware devices and software algorithms transform digital representation of a virtual environment into signals for displays, which then present images in a way that can be perceived by human senses. Different senses require different stimuli, so rendering is usually performed using different hardware and software platforms.

Presentation of information in virtual reality allows freedom of expression, but also establishes certain restrictions. Most restrictions result from the difficulty of implementation of rendering, which needs to be done in real time, but must enable stereoscopic visual information, spatial audio information and haptic information of various modalities. In turn, virtual reality enables interactive motion in space and the ability to manipulate objects in a manner that is similar to normal handling of objects in a real environment.

Although the aim is to create a uniform virtual environment for all senses, the details of implementation vary and will be specifically addressed for visual, acoustic and haptic rendering.

1.3.2.1 Visual Rendering

Visual rendering is a research area addressed by computer graphics. Rendering can be based on geometric or non-geometric methods. Geometric surface rendering is based on the use of polygons, implicit and parametric surfaces and constructive solid geometry. The polygonal method is the simplest and can, with a partial loss of information, be used to display objects modeled using implicit and parametric surfaces. Polygons are plane shapes bounded by a closed path composed of at least three line segments. In visual rendering three- or four-sided polygons are usually used for performance reasons. Parametric and implicit surfaces enable description of curved objects. Constructive solid geometry allows a modeler to create a complex

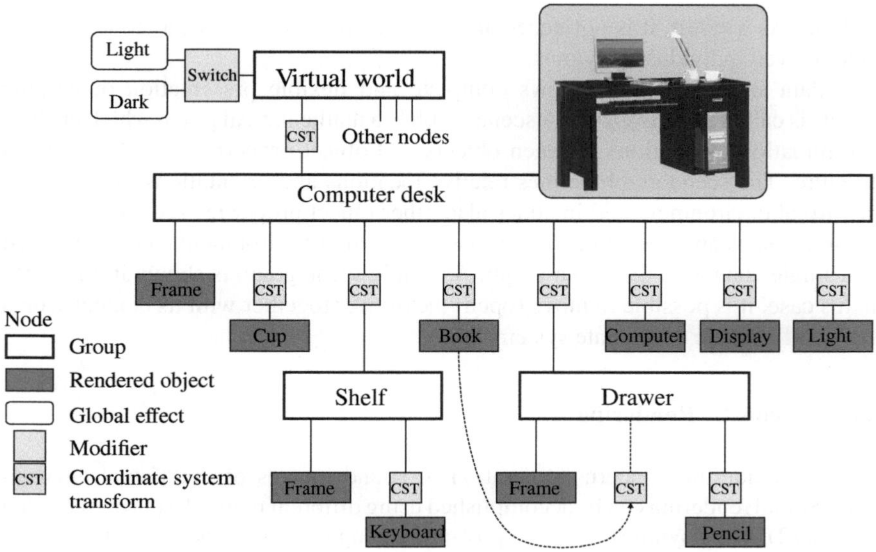

Fig. 1.10 The scene graph enables grouping of dependent objects in order to simplify definition of their parameters

surface or object by using Boolean operators (union, intersection or difference) to combine object primitives (cuboids, cylinders, prisms, pyramids, spheres, cones).

Methods based on object surface modeling are most appropriate for description of opaque objects. The use of geometric methods is problematic in case of transparent objects. This is particularly the case for spaces that are filled with variable-density semitransparent substances. Non-geometric (non-surface) rendering of objects includes the volumetric method and methods based on particle description. Volumetric rendering is appropriate for semitransparent objects and is often used for presentation of medical, seismic and other research data. It is based on the ray-tracing method—a technique for generating an image by tracing the path of light through pixels in an image plane. Light rays, which are subjected to laws of optics, are altered due to reflections from surfaces describing virtual objects. Their properties are also altered when passing through semitransparent material.

Generation of visual scenes requires adequate presentation of the form and pose of virtual objects. Polygons are the most common way of representing objects and computer graphic cards are usually optimized for polygon rendering. In addition to positions of polygon vertices, it is also necessary to specify color, texture and surface parameters that are related to individual polygons. In order to simplify representation of objects, polygons must be grouped into simple geometric forms such as cubes, spheres or cones, which are then only decomposed into polygons in the graphic card processing unit. Similarly, it is possible to group polygons into sets which represent objects such as a chair or a table. Grouping allows easy positioning of the object as a whole. As a result, it is not necessary, for example, to move separate legs of the table or even individual polygons.

A data structure which allows complete and flexible presentation of graphic objects is called a *scene graph*. A scene graph is a mathematical graph which enables determination of relations between objects and object properties in a hierarchical structure. The scene graph defines relative locations and orientations of objects in the virtual environment and includes also other object properties such as color and texture. A substantial modification of part of a virtual environment can be triggered by a single change in the scene graph. A sample scene graph is shown in Fig. 1.10. In this case, it is possible to move (open) the drawer together with its content with a change of a single coordinate system.

1.3.2.2 Acoustic Rendering

Acoustic rendering concerns generation of sound images of the virtual environment. Sound rendering can be accomplished using different methods: (1) prerecorded sounds, (2) sound synthesis and (3) post-processing of prerecorded sounds.

A common method of generating sound is by playing prerecorded sounds from the real environment. This is particularly suitable for generating realistic audio presentation. Several sampled sounds can be merged or modified, making it possible to produce a more abundant and less repetitive sound.

Sound synthesis based on computer algorithms allows greater flexibility in generating sound, but makes it harder to render realistic sounds. Sound synthesis is based on spectral methods, physical models or abstract synthesis. Physical modeling allows generation of sounds using models of physical phenomena. Such sounds can be very realistic. Sounds can simulate continuous or discrete events such as, for example, sounds of colliding objects.

Postprocessing—recorded sounds or sounds generated in real time can be additionally processed, which results in sounds similar to the original, but with certain qualitative differences. Added effects may be very simple such as, for instance, an echo that illustrates that the sound comes from a large space, or attenuation of high frequency sounds, which results in an impression of distance of sound origin.

1.3.2.3 Haptic Rendering

Rendering of haptic cues often represents the most challenging problem in a virtual reality system. The reason is primarily the direct physical interaction and, therefore, a bidirectional communication between the user and the virtual environment through a haptic display. The haptic interface is a device that enables man-machine interaction. It simultaneously generates and perceives mechanical stimuli.

Haptic rendering allows the user to perceive the mechanical impedance, shape, texture and temperature of objects. When pressing on an object, the object deforms due to its final stiffness, or moves, if it is not grounded. The haptic rendering method must take into account the fact that humans simultaneously perceive tactile as well as kinesthetic cues. Due to the complexity of displaying tactile and kinesthetic cues, virtual reality systems are usually limited to only one type of cue. Haptic rendering can thus be divided into rendering through the skin (temperature and texture) and rendering through muscles, tendons and joints (position, velocity, acceleration, force and impedance).

Stimuli, which trigger mainly skin receptors (e.g. temperature, pressure, electrical stimuli and surface texture), are displayed through tactile displays. Kinesthetic information that enables the user to investigate object properties such as shape, impedance (stiffness, damping, inertia), weight and mobility, are usually displayed through robot-based haptic displays. Haptic rendering can produce different kinds of stimuli, ranging from heat to vibrations, movement and force. Each of these stimuli must be rendered in a specific way and displayed through a specific display.

Temperature rendering is based on heat transfer between the display and the skin. The tactile display creates a sense of object temperature.

Texture rendering provides tactile information and can be achieved, for example, using a field of needles, which simulate the surface texture of an object. Needles are active and adapt according to the current texture of the object being explored by the user.

Kinesthetic rendering allows display of kinesthetic information and is usually based on the use of robots. By moving the robot end-effector, the user is able to haptically explore his surroundings and perceive the position of an object, which is determined by an inability to penetrate the space occupied by that object. The greater the stiffness of the virtual object, the stiffer the robot manipulator becomes while in contact with a virtual object. Kinesthetic rendering thus enables perception of the object's mechanical impedance.

Haptic rendering of a complex scene is much more challenging compared to visual rendering of the same scene. Therefore, haptic rendering is often limited to simple virtual environments. The complexity of haptic rendering arises from the need for a high sampling frequency in order to provide consistent feeling of rendered objects. If the sampling frequency is low, the time required for the system to respond and produce an adequate stiffness (for example, during penetration into a virtual object) becomes noticeable. Consequently, stiff objects feel compliant.

The complexity of realistic haptic rendering depends on the type of simulated physical contact implemented in the virtual reality. If only the shape of an object is being displayed, then *touching* the virtual environment with a pencil-style probe is sufficient. Substantially more information needs to be transmitted to the user if it is necessary to grasp the object and raise it to feel its weight, elasticity and texture. Therefore, the form of the user contact with the virtual object needs to be taken into account for the haptic rendering (for example, contact can occur at a single point, the object can be grasped with the entire hand or with a pinch grip between two fingers).

Single-point contact is the most common method of interaction with virtual objects. The force display provides stimuli to a fingertip or a probe that the user holds with his fingers. The probe is usually attached as a tool at the tip of the haptic interface. In the case of the single point contact, rendering is usually limited to the contact forces only and not contact torques.

Two-point contact (pinch grip) enables display of contact torques through the force display. With a combination of two displays with three degrees of freedom it is, in addition to contact forces, possible to simulate torques around the center point on the line, which connects the points of touch.

Multipoint contact allows object manipulation with six degrees of freedom. The user is able to modify both the position and the orientation of the manipulated object. To ensure adequate haptic information, it is necessary to use a device that covers the entire hand (a haptic glove).

As with visual and acoustic rendering, the amount of details or information that can be displayed with haptic rendering is limited. Usually the entire environment is required to be displayed in a haptic form. However, due to the complexity of haptic rendering algorithms and specificity of haptic sensing, which is local in nature, haptic interactions are often limited to contact between the probe and a small number of nearby objects. Due to the large amount of information necessary for proper representation of object surfaces and dynamic properties of the environment, haptic rendering requires a more detailed model of a virtual environment (object dimensions, shape and mechanical impedance, texture, temperature), than is required for visual or acoustic rendering. Additionally, haptic rendering is computationally more demanding than visual rendering, since it requires accurate computation of contacts between objects or contacts between objects and tools or avatars. These contacts form the basis for determining reaction forces.

1.4 Display Technologies

The virtual reality experience is based on the user's perception of the virtual environment. Physical perception of the environment is based on a computer display. The concept of *display* represents all methods of presenting information to any human sense. The human sensory system integrates various senses that provide information about the external environment to the brain. Three of these senses—vision, hearing and touch—are most often used in presentation of synthetic stimuli originating from a virtual reality system. The system *fools* human senses by displaying computer-generated stimuli which replace or augment natural stimuli available to one or more type of receptors. A general rule is that the higher the number of senses being excited by synthetic stimuli, the better the virtual reality experience.

In principle, displays can be divided into three major categories: stationary (grounded), attached to the body as exoskeletons, or head-mounted. Stationary displays (projection screens, speakers) are fixed in place. In a virtual reality system, their output is adjusted in a way that reflects changes in position and orientation of the user's senses. Tracking of user's pose in space is required. Head-mounted displays move together with the user's head. Consequently, the display keeps constant orientation relative to the user's senses (eyes, ears), independently of head orientation. Displays attached to the user's limbs (most often arm or hand) move together with the respective limb.

1.4.1 Visual Displays

Visual displays are optimized to correspond to the characteristics of human vision apparatus. Though all visual displays present a visual image to the user, they differ in a number of properties. These properties determine quality of visual presentation of information and affect user's *mobility* within the system. User's mobility may have an impact on immersion as well as on the usefulness of virtual reality application. Most displays provide restrictions on mobility, which are the result of limitations of movement tracking devices, electrical connections or the fact that the display is stationary.

Two visual channels are required to produce stereoscopic images, with each of the two channels presenting an image for one eye. Different multiplexing methods may be used to separate images for the left eye and right eye: *spatial multiplexing* requires separate display for each eye, *temporal multiplexing* requires a single display with-time multiplexed images and active shutter glasses synchronized with the display (Fig. 1.11), *spectral multiplexing* is based on presenting images in different parts of the visible light spectrum for the left and right eye and the use of colored glasses which separate the two images, *light polarization technology* is based on linear or

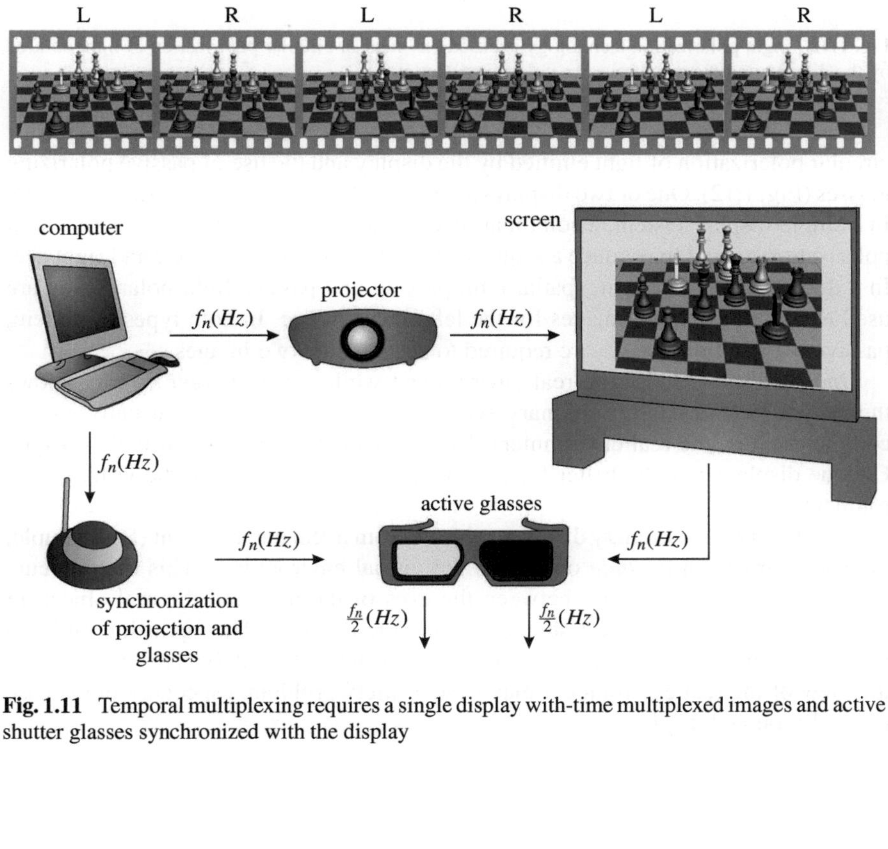

Fig. 1.11 Temporal multiplexing requires a single display with-time multiplexed images and active shutter glasses synchronized with the display

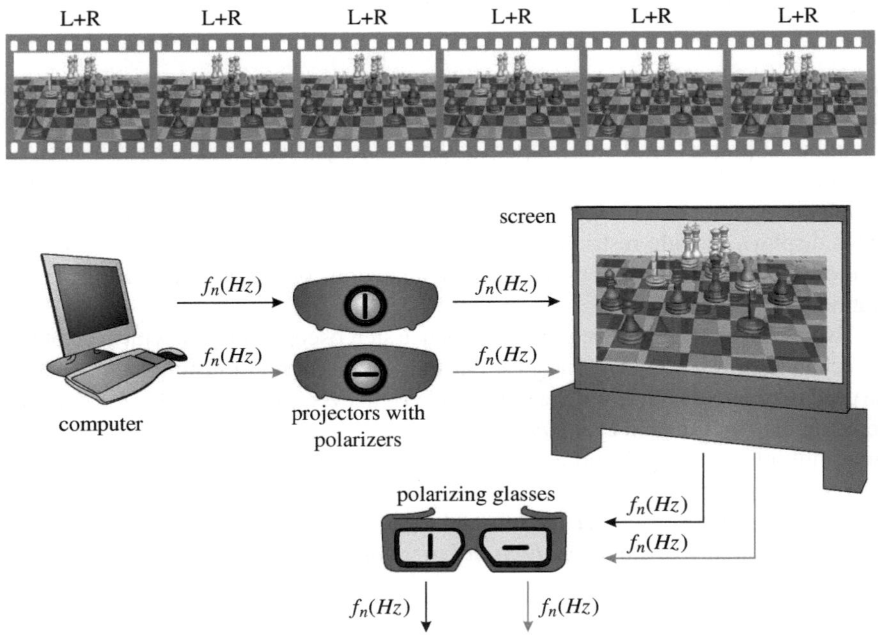

Fig. 1.12 Light polarization technology is based on linear or circular polarization of light emitted by the display and the use of passive polarizing glasses

circular polarization of light emitted by the display and the use of passive polarizing glasses (Fig. 1.12). One or two displays are required to produce a stereoscopic image. In a single-display system, a combination of temporal multiplexing and active light polarization is used to produce a sequence of polarized images for left and right eye. In a double-display system, spatial multiplexing and passive light polarization are used to produce parallel images for the left and right eye. In both types of system, passive polarization glasses are required to separate the two images.

Opaque display hides the real environment while a *transparent display* allows the user to see through. Stationary screens and desktop displays usually cannot completely hide the real environment. Head-mounted displays are usually opaque. Opaque displays may be better for achieving user's immersion in the virtual environment.

With the use of stationary displays, objects from a real environment (for example, the user's arm) can *occlude* objects in the virtual environment. This often occurs when a virtual object comes between the eyes of the user and the real object, in which case the virtual object should occlude the real one. The occlusion problem is less pronounced when using head-mounted displays. Visual displays that occlude the view of the real environment may pose a safety problem, especially in the case of head-mounted displays.

Field of view is the angular extent of the observable world that is seen at any given moment. Binocular vision, which is important for depth perception, only covers 140° of the field of vision in humans. The remaining peripheral 40° have no binocular vision (because of the lack of overlap in the images from either eye for those parts of the field of view). The field of view of a visual display is the measure of the angular width of the user's field of view, which is covered by the display at any given time. For example, the field of view of head-mounted displays determines the extent to which the operator can see visual information without moving his or her head. The *field of regard* refers to the area within which the operator can move his or her head to see visual information.

Due to the complexity of the virtual environment rendering, a significant time delay between the movement of the user and an adequate response seen through the display may occur. Similar effects occur if frame rate with which images are refreshed is too low. Refresh frequency depends mainly on the hardware equipment used for rendering a virtual environment and on the computing complexity of the virtual environment. Such delays may cause discomfort for the user. Permissible visual display latency in an augmented reality system is considerably smaller than for the classical virtual reality system, since long delays cause desynchronization between the real-world and the computer-generated visual stimuli.

Visual displays can be categorized based on various properties. A general categorization that emphasizes size and mobility of displays results in four distinct display categories.

Desktop display is the simplest display based on a computer screen with or without stereoscopic vision. The display shows a three-dimensional image which varies depending on the location of the user's head, thus it is necessary to track head movements.

A projection-based display screen size is usually much larger than a desktop display and, therefore, covers a large part of the user's field of view. Projection screens can encircle the user and thereby increase the field of regard. The larger display size allows the user to walk within the limited area in front of the display. The size of projection-based displays defines requirements for tracking devices that detect user movements. A stereoscopic effect can be achieved by any type of image multiplexing. In contrast to head-mounted displays, the user is not isolated from the real environment.

Head-mounted display (Fig. 1.13) can be transparent or opaque. Head-mounted displays are portable and move together with the head of the user. Screens of head-mounted displays are usually small and light and allow stereoscopic vision. A common drawback of head-mounted displays is the delay between the movement of the head and the change of the displayed image. The field of view of typical head-mounted displays is usually quite limited while the field of regard is large since the display is constantly positioned in front of the user's eyes. Non-see-through head-mounted displays hide the real environment, therefore, everything that a user needs to see must be artificially generated—also the representation of the user himself if required. See-through head-mounted displays are primarily intended for augmented reality applications. Transparency of display can be achieved by using lenses and

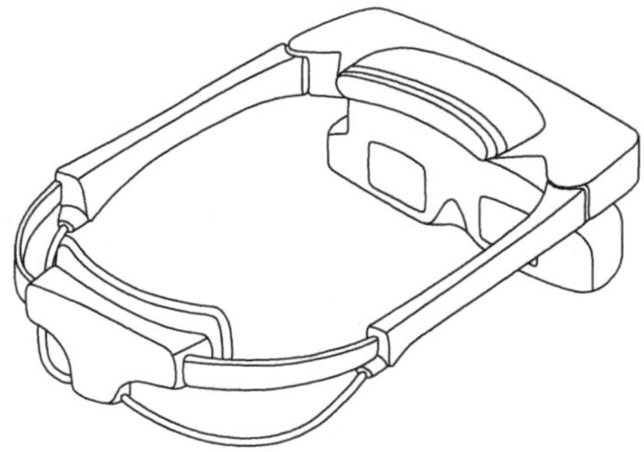

Fig. 1.13 Head mounted display

semi-transparent mirrors or by using a video method that superimposes the virtual reality image over the real environment video image. In augmented reality systems, the real environment is part of the scenario, thus, limitations of the real environment affect the characteristics of the virtual environment. Given that augmented reality allows an indirect view of the real world, it is easy to use it as an interface to a geographically remote environment, which leads to telepresence.

A hand-held display consists of a small screen that the user holds in his hand. The image on the screen is adjusted according to changes in orientation of the vector between the screen and the eyes of the user. These displays are most often used for augmented-reality applications.

1.4.2 Auditory Displays

Auditory displays generate sound to communicate information from a computer to the user. Similarly to visual displays, acoustic displays can also be divided into two major categories: fixed and head-mounted displays. Headphones represent an analogy to visual head-mounted displays and can either completely separate the user from the sounds of the real environment or allow real environment sounds to overlap with the artificial stimuli.

The same as eyes in relation to visual display, also ears can be presented with the same information (monophonic display) or different information (stereophonic display). Due to the interaural distance, ears generally perceive slightly different information—the same signal travels different paths before it reaches each ear. These different pathways help the brain to determine the origin of sound. Use of stereophonic headphones enables rendering of such cues. If sound localization is not correlated with localization of visual information, the combined effect can be very annoying.

In a real environment, humans perceive three-dimensional characteristics of sound through various sound cues. Human brains are generally able to localize the source of sound by combining the multitude of cues that include interaural time delay (the difference in arrival time of the same audible signal between two ears), the difference of the signal amplitude in each ear, echoes, reflected sounds, filtering of sound resulting from the sound passing through various materials, absorption of certain frequencies in the body and filtering through the external ear.

In principle, audio displays can be divided into two types: head-mounted displays (headphones) and stationary displays (loudspeakers). Headphones that move along with the head of the user, are intended for one user and allow implementation of an isolated virtual environment. Similarly to visual head-mounted displays, headphones can isolate the user from the real environment sounds or allow synthetic sounds to mix with the real environment stimuli through open headphones. Since headphones provide two-channel stimulation, it is in principle possible to simulate three-dimensional sound more easily than with loudspeakers. In general, headphones display sound that is computed based on the orientation of the head. If the sound originates from a point in space, it is necessary to track the movement of the head and appropriately compute the synthesized sound.

Speakers are better suited for use with projection-based visual displays. The stationary nature of speakers results in audio defined in relation to the environment. This allows generation of sound independently of the position and orientation of the user's head and allows greater mobility of the user. Since speakers create a combination of direct and reflected sound, they make it more difficult to control the sound that comes to each ear. With headphones, the user hears only the direct sound, thus, information can be presented in greater detail.

1.4.3 Haptic Displays

The sense of touch is often used to verify object existence and mechanical properties. Since haptic sense is difficult to deceive, implementation of a haptic display is very challenging. The concept of haptics refers to the kinesthetic and tactile senses. Kinesthesiology represents perception of movement or tension in muscles, tendons and joints. Tactile perception arises from receptors in the surface of the skin. Tactile perception includes temperature, pressure and forces within the area of contact.

In general, the use of haptic displays in virtual reality applications is less frequent than the use of visual and sound displays. Haptic displays are more difficult to implement than visual or audio displays due to the bidirectional nature of the haptic system. Haptic displays do not only enable perception of the environment, but also manipulation of objects in the environment. Thus, the display requires direct contact with the user. Since active haptic feedback can be difficult to implement, it is sufficient to use passive haptic feedback in certain applications. In this case, the display does not generate active forces as a reaction to the user's actions. Instead, real objects are used as components of the user interface in order to provide information about the virtual environment. The easiest way to implement such passive haptic feedback is by using control props.

While haptic displays are analyzed in great detail in Chaps. 2–11, we will summarize some of their basic properties here to complete the introduction to virtual reality systems. Haptic properties determine the quality of the virtual reality experience.

Kinesthetic cues represent a combination of sensory signals that enable awareness about joint angles as well as muscle length and tension in tendons. They allow the brain to perceive body posture and the environment around us. The human body consists of 75 joints (44 of them in the hands) and all joints have receptors that provide kinesthetic information, therefore, it becomes impossible to cover all possible points of contact with the body with a single haptic display.

To reach any pose in three-dimensional space, a display with six *degrees of freedom* is required. In general, haptic displays can be found with any number of degrees of freedom (most often up to six). Displays with less than four degrees of freedom are usually limited to rendering position and force.

Force displays require a *grounding point*, which provides support against the forces applied by the user. Grounding can be done relative to the environment or to the user. Haptic displays that are grounded to the environment restrict the mobility of the user. On the other hand, displays that are grounded to the user allow users to move freely in a large space.

Tactile cues represent a combination of sensory signals from receptors in the skin that collect information about their close proximity. Mechanoreceptors enable collection of accurate information about the shape and surface texture of objects. Thermoreceptors perceive heat flow between the object and the skin and pain receptors perceive pain due to skin deformation or damage.

The ability of a human sensory system to distinguish between two different nearby tactile stimuli varies for different parts of the body. This information defines the required spatial resolution of a haptic display that must be, for example, higher for the fingertips than for the skin on the upper arm.

Haptic displays may exist in the form of desktop devices, exoskeleton robots or large systems, which can move greater loads. Given the diversity of haptic feedback (tactile, proprioceptive, and thermal) and the different parts of the body to which the display can be coupled, display mechanisms are usually highly optimized for specific applications (Fig. 1.14).

Design of haptic displays requires compromises which ultimately determine the realism of virtual objects. Realism defines how realistically certain object properties (stiffness, texture) can be displayed compared to direct contact with a real object. Low refresh rate of a haptic interface, for example, significantly deteriorates the impression of simulated objects. Objects generally feel softer and contact with objects results in annoying vibrations which affect the feeling of immersion. A long delay between an event in a virtual environment and responses of the haptic display furthermore degrades the feeling of immersion. Since haptic interactions usually require hand-eye coordination, it is necessary to reduce both visual as well as haptic latencies and synchronize both displays.

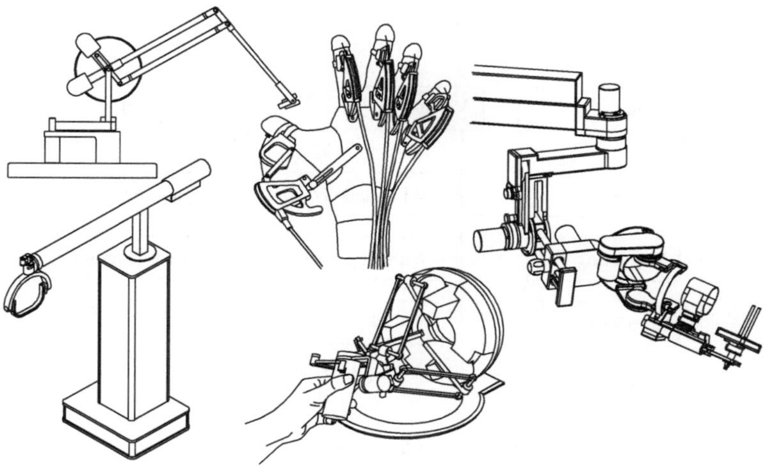

Fig. 1.14 A collage of different haptic robots for upper extremities: Phantom (Sensable), Omega (Force Dimension), HapticMaster (Moog FCS), ARMin (ETH Zurich) and CyberGrasp (CyberGlove Systems)

Safety is of utmost importance in dealing with haptic displays in the form of robots. High forces, which may be generated by haptic devices, can damage the user in case of a system malfunction.

1.4.4 Vestibular Displays

The vestibular sense enables control of balance. The vestibular receptor is located in the inner ear. It senses acceleration and orientation of the head in relation to the gravity vector. The relation between vestibular sense and vision is very strong and the discrepancy between the two inputs can lead to nausea.

The vestibular display is based on the physical movement of the user. A movement platform can move the ground or the seat of the user. Such platforms are typical in flight simulators. A vestibular display alone cannot generate a convincing experience, but can be very effective in combination with visual and audio displays.

1.5 Input Devices to Virtual Reality System

A virtual reality system allows different modes of communication or interaction between the user and the virtual environment. In order to enable immersion of a user in a synthetic environment, a device that detects the location and actions of the

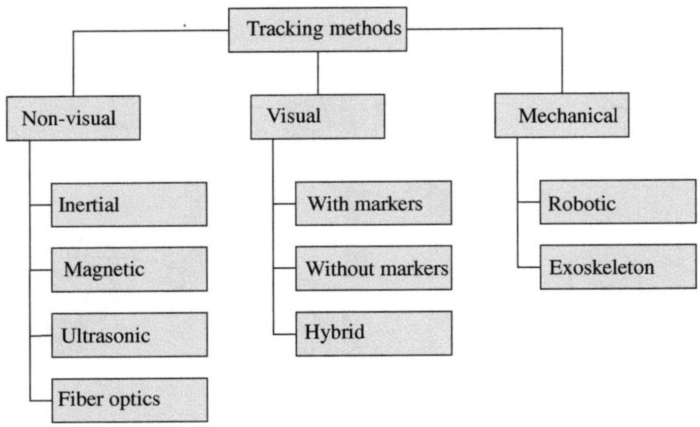

Fig. 1.15 Different methods for user motion tracking

user is required. Continuous user movement tracking allows the system to display the virtual environment from the correct user's perspective, which is a prerequisite for establishment of the physical immersion. Input signals generated by the user and acquired by the virtual reality system allow interaction with the virtual environment. User interaction with a virtual environment through a virtual reality system enables bidirectional exchange of information through various input and output devices.

User movement tracking is one of the basic components of a virtual reality system. Movement tracking is possible using active methods (spoken commands, different platforms as well as controllers such as joysticks, keyboards, or steering wheels), which allow the user to directly input the information into the virtual reality system, as well as using passive methods, which measure user movement and provide information about the user movement and gaze direction to the computer. In addition to tracking the user, it is often necessary also to perceive the user's surrounding. This allows display of information from the real environment augmented with synthetic stimuli.

1.5.1 Pose Measuring Principles

Pose tracking allows measurement of the user's position and orientation in space. A pose sensor is a device that enables measurement of an object pose. It is one of the most important measurement devices in a virtual reality system. Methods for pose detection are based on various principles (Fig. 1.15).

The electromagnetic principle requires a transmitter with three orthogonal coils that generate a weak magnetic field, which then induces currents in receiver coils (Fig. 1.16). By measuring the currents in the receiver, it is possible to determine the relative position and orientation between the transmitter and the receiver. The receiver

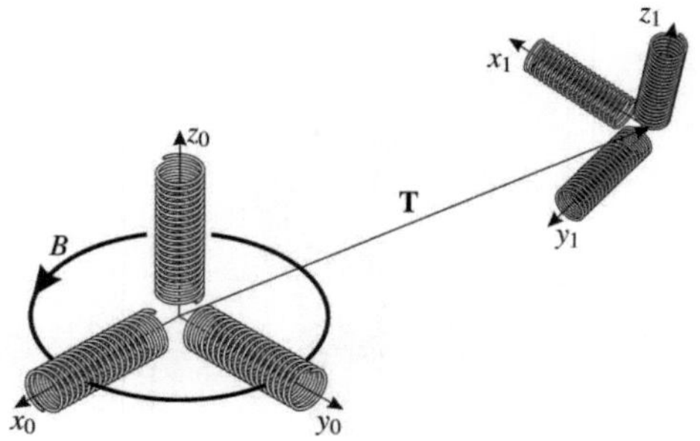

Fig. 1.16 Electromagnetic transmitter and receiver coils for computation of relative pose **T**

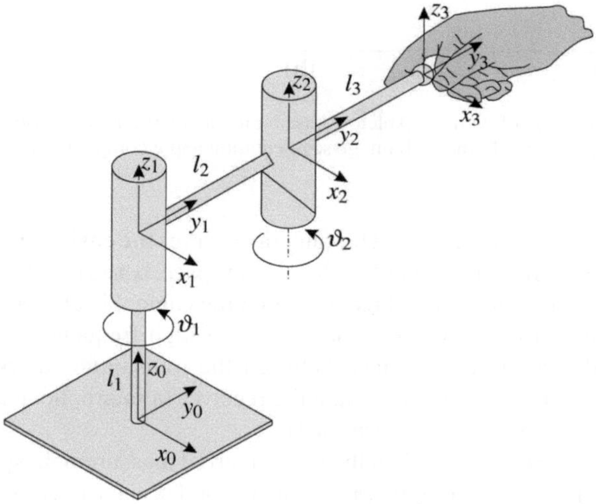

Fig. 1.17 An example of a mechanism with two degrees of freedom in contact with a human hand

signal depends both on the receiver's distance from the transmitter as well as their relative orientation. The system allows measuring the pose of the receiver in six degrees of freedom.

The mechanical principle is based on a mechanism with multiple degrees of freedom that is equipped with joint position sensors. The mechanism is physically connected to the user (Fig. 1.17). The device tracks user movements. To increase mechanism ergonomics, a weight compensation system may be implemented to compensate for the mechanism weight.

The optical principle uses visual information to detect user movement. Measurements can be accomplished using video cameras or dedicated cameras with active or passive markers. Computation of a human skeleton based on the markerless optical motion tracking technology is shown in Fig. 1.18. A special case is the videometric principle where the camera is not fixed in space, but rather attached to the object

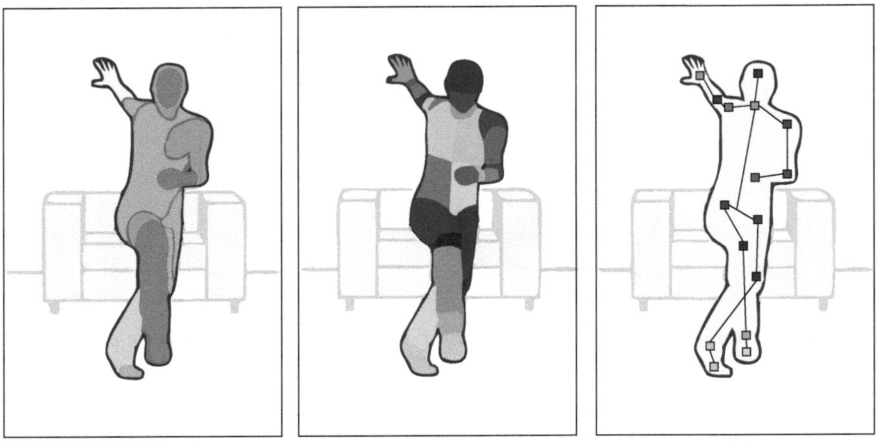

(a) (b) (c)

Fig. 1.18 Computation of human skeleton based on the markerless optical motion tracking technology; **a** acquired depth image, **b** image segmentation and **c** computation of body skeleton

whose location is being measured. The camera observes the environment. The videometric principle requires use of markers located in space, based on which it is possible to determine the location of the object to which the camera is affixed.

The ultrasonic principle is based on the use of high-frequency sounds, making it possible to determine the distance between the transmitter (a speaker, usually attached in a fixed location in space) and the receiver (a microphone attached to the object whose location is being determined).

The inertial principle is based on the use of inertial measurement systems consisting of a triad of gyroscopes (angular rate sensors) and accelerometers. The system is often augmented with magnetometers (usually measuring their relative orientation with respect to the Earth's magnetic field). Inertial tracking is similar to perceptions of the human inner ear, which estimates head orientation. In principle, an inertial measurement unit allows measurement of six degrees of freedom that define the pose of the measured object. However, technical challenges exist that make position measurements unreliable. The errors are the result of non-ideal outputs of acceleration sensors (bias, drift). A concept of sensory fusion for inertial tracking is shown in Fig. 1.19. Inertial sensors are often installed in head-mounted displays and allow detection of movement and pose of the user's head.

1.5.2 Tracking of User Pose and Movement

Tracking of user pose and movement allows detection of user pose and actions in a virtual reality system. The virtual reality application determines which body movements (which segments) need to be measured. The term *gesture* describes a specific

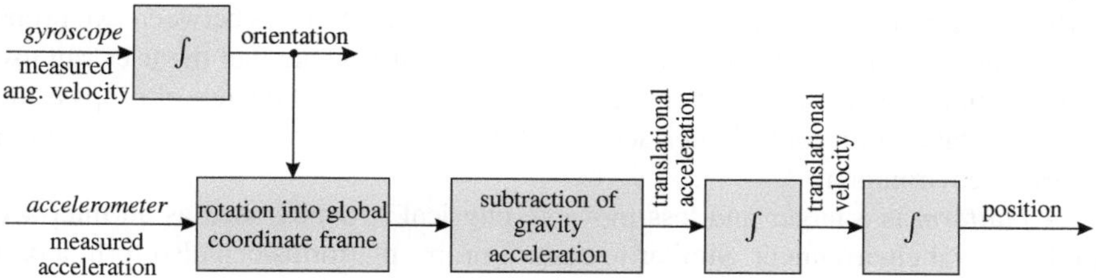

Fig. 1.19 Inertial tracking concept

movement that occurs at a given time. Gestures allow intuitive interaction with the virtual environment. Different parts of the human body can be tracked and their movements used as inputs to the virtual reality system.

Head movement tracking is necessary for proper selection of perspective in a virtual environment. With a stationary display it is important to determine the position of the viewer's eyes with respect to the display. For a correct stereoscopic presentation, it is necessary to track all six degrees of freedom of the head. When using a head-mounted display, information about head orientation is important for proper presentation of visual information. The displayed content depends on the relative rotation of the head (for example, when the user turns his head to the left, he expects to see what is on his left side). *Eye tracking* can be combined with head movement tracking in order to obtain gaze direction.

Arm, hand and finger movement tracking allows the user to interact with a virtual environment. In multiuser environments gesture detection allows communication between users through their respective avatars.

Torso movement tracking provides better information on the direction of movement of the body compared to information obtained from head orientation. If head orientation is used to determine the movement direction, the user cannot turn his head to see sideways, while walking straight ahead.

1.5.3 Physical Input Devices

Physical input devices are typically a part of the interface between the user and the virtual environment and are either simple objects, which are held in the hand, or complex platforms; the person that operates the physical device gets a certain sense of the object's physical properties, such as weight and texture, which represents a type of haptic feedback.

Physical control inputs can be individual buttons, switches, dials or sliders, allowing direct input into the virtual reality system.

A control prop is a physical object used as an interface to a virtual environment. Physical properties of a prop (shape, weight, texture, hardness) usually imply its use in a virtual environment. A prop allows intuitive and flexible interaction with the virtual environment. The ability to determine the spatial relations between two props or between a prop and the user provides a strong sensory cue that the user can use to better understand the virtual environment. The objective of the use of props is the implementation of a control interface that allows user's natural manipulation within a virtual environment.

A platform is a larger and less movable physical structure used as an interface to the virtual environment. Similar to control props, platforms can also form a part of the virtual environment by using real objects with which the user can interact. Such a platform becomes a part of the virtual reality system. A platform can be designed to replicate a device from the real environment, which also exists in the virtual environment. An example of a platform may be the cockpit of an airplane.

1.6 Interaction with a Virtual Environment

Interaction with a virtual environment is the most important feature of virtual reality. Interaction with a computer generated environment requires the computer to respond to the user's actions. The mode of interaction with the computer is determined by the type of the user interface. Proper design of the user interface is of utmost importance since it must guarantee the most natural interaction possible. The concept of an ideal user interface uses interactions from the real environment as metaphors through which the user communicates with the virtual environment.

Interaction with a virtual environment can be roughly divided into *manipulation*, *navigation* and *communication*. Manipulation allows the user to modify the virtual environment and to manipulate objects within it. Navigation allows the user to move through the virtual environment. Communication can take place between different users or between users and intermediaries in a virtual environment.

1.6.1 Manipulation Within the Virtual Environment

One of the advantages of an interactive virtual environment is the ability to interact with objects or to manipulate them in this environment. Some manipulation methods are shown in Fig. 1.20.

Direct user control allows a user to interactively manipulate an object in a virtual environment the same way as he would in the real environment.

Physical control enables manipulation of objects in a virtual environment with real environment devices (buttons, switches, haptic robots). Physical control allows passive or active haptic feedback.

Virtual control allows manipulation of objects through computer-simulated devices (simulation of real world devices—virtual buttons, steering wheel) or avatars (intelligent virtual agents). The user activates a virtual device via an interface (real device), or commands can be sent to an avatar that performs the required action.

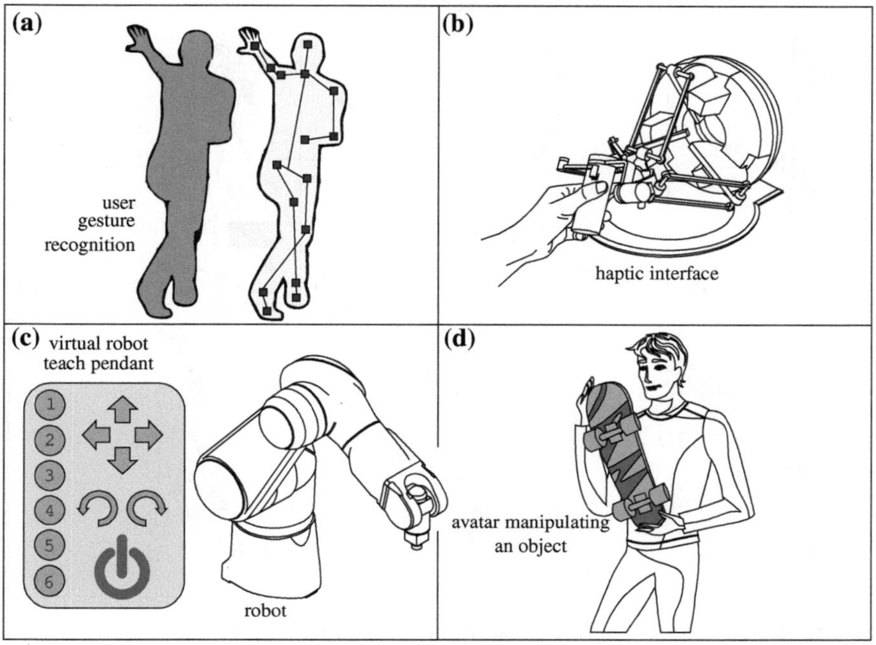

Fig. 1.20 Manipulation methods: **a** direct user control (gesture recognition), **b** physical control (buttons, switches, haptic robots), **c** virtual control (computer-simulated control devices) and **d** manipulation via intelligent virtual agents)

The advantage of a virtual control is that one real device (for example a haptic robot) activates several virtual devices.

Manipulation of location and shape of objects allows changing the position and orientation of an object and its shape by one of the methods of manipulation. *Application of force on a virtual object* allows interactions such as object grasping, pushing, squeezing and hitting. A haptic robot allows a realistic presentation of forces acting on the object being manipulated. *Modification of the state of virtual control interfaces* allows the position of switches, buttons, sliders and other control functions implemented in a virtual environment to be changed. *Modification of object properties* allows quantities such as transparency, color, mass, density of objects to be changed. These are operations that do not replicate real-environment actions.

Manipulation in a virtual environment is usually based on similar operations as in the real environment; however, interfaces and methods exist that are specific. *Feedback* may exist in visual, aural or haptic form. However, one type of information can often be substituted with another (for example, haptic information can be substituted with audio cues). On the other hand, *virtual fixtures*, for example, allow easier and safer execution of tasks in a virtual environment (motion of an object, for example, can be limited along a single axis).

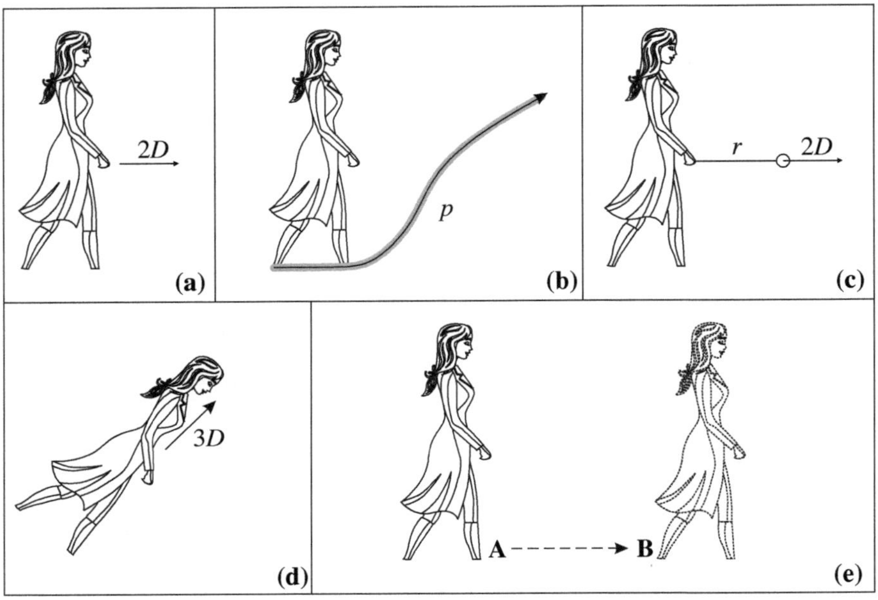

Fig. 1.21 Traveling methods: **a** locomotion, **b** path tracking, **c** towrope, **d** flying and **e** displacement

1.6.2 Navigation Within the Virtual Environment

Navigation represents movement in space from one point to another. It includes two important components: (1) *travel* (how the user moves through space and time) and (2) *path planning* (methods for determination and maintenance of awareness of position in space and time, as well as trajectory planning through space to the desired location). Knowing the location and neighborhood is defined as position awareness.

In a virtual environment, where the area of interest extends beyond direct virtual reach of the user, traveling is one possibility for space exploration. Some traveling methods are shown in Fig. 1.21.

Physical locomotion is the simplest way to travel. It requires only tracking the user's body movement and adequate rendering of the virtual environment. The ability to move in real space also enables proprioceptive feedback, which helps to create a sense of relationships between objects in space. A device that tracks user movement must have a sufficiently large working area. *Path tracking or a virtual tunnel* allows the user to follow a predefined path in a virtual environment. The user is able to look around, but cannot leave the path. *The towrope method* is less constraining for the user than path tracking. The user is towed through space and may move around the coupling entity in a limited area. *Flying* does not constrain the user movement to a surface. It allows free movement in three-dimensional space. At the same time it enables a different perspective of the virtual environment. The fastest way of moving through a virtual environment is a simple *displacement* that enables movement between two points without navigation (the new location is reached in a time instant).

1.6.3 Interaction with Other Users

Simultaneous operation of more users in a virtual environment is an important property of virtual reality. Users' actions in virtual reality can be performed in different ways. If users work together in order to solve common problems, the interaction results in *cooperation*. However, users may also compete among themselves or interact in other ways. In an environment where many users operate at the same time, different issues need to be taken into account. It is necessary to specify how interaction between persons will take place, who will have control over the manipulation or communication, how to maintain the integrity of the environment and how the users communicate. Communication is usually limited to visual and audio modality. However, it can also be augmented with haptics.

Certain professions require cooperation between experts to accomplish a task within a specified time. In addition to manipulation tasks, where the need for physical power requires the cooperation of several persons, there are many tasks that require cooperation between experts such as architects, researchers or medical specialists. The degree of participation in a virtual environment may extend from zero, where users merely coexist in a virtual environment, to the use of special tools that allow users to simultaneously work on the same problem.

Interactive cooperation requires environmental coherency. This defines the extent to which the virtual environment is the same for all users. In a completely coherent environment, any user can see everything that other users do. Often it is not necessary for all features of the virtual environment to be coherent. Coherency is of primary importance for simultaneous cooperation, for example, when more users work on a single object.

Reference

1. Sherman, W.R., Craig, A.B.: Understanding virtual reality. Morgen Kaufman Publishers, San Francisco (2003)

Chapter 2
Introduction to Haptics

2.1 Definition of Haptics

The word *haptic* originates from the Greek verb *hapto—to touch*—and therefore refers to the ability to touch and manipulate objects. The haptic experience is based on tactile senses, which provide awareness of the stimuli on the surface of the body and kinesthetic senses, which provide information about body pose and movement. Its bidirectional nature is the most prominent feature of haptic interaction, which enables exchange of (mechanical) energy—and therefore information—between the body and the outside world.

The word *display* usually emphasizes the unidirectional nature of transfer of information. Nevertheless, in relation to haptic interaction, similar to visual and audio displays, the phrase *haptic display* refers to a mechanical device for transfer of kinesthetic or tactile stimuli to the user.

The term haptics often refers to sensing and manipulation of virtual objects in a computer-generated environment—a synthetic environment that interacts with a human when performing sensory-motor tasks. A typical virtual reality system consists of a head-mounted display, which projects computer-generated images and sound based on the user's head orientation and gaze direction and a haptic device that allows interaction with a computer through gestures. Synthesis of virtual objects requires an optimal balance between the user's ability to detect object's haptic properties, the computational complexity required to render objects in real time and the accuracy of haptic devices for generating mechanical stimuli.

Virtual environments that engage only the user's visual and auditory senses are limited in their ability to interact with the user. It is desirable to also include a haptic system that not only transmits sensations of contact and properties of objects, but also allows their manipulation. The human arm and hand enable pushing, grasping, squeezing, or hitting the objects, they enable exploration of object properties such as surface texture, shape and compliance and they enable manipulation of tools such as a pen or a hammer. The ability to touch, feel and manipulate objects in a virtual environment, augmented with visual and auditory perception, enables a degree of

M. Mihelj and J. Podobnik, *Haptics for Virtual Reality and Teleoperation*, Intelligent Systems, Control and Automation: Science and Engineering 64,
DOI: 10.1007/978-94-007-5718-9_2,

immersion that otherwise would not have been possible. The inability to touch and feel objects, either in a real or a virtual environment, impoverishes and significantly affects the human ability of interaction with the environment [1].

A haptic interface is a device that enables interaction with virtual or physically remote environments [2, 3]. It is used for tasks that are usually performed by hand in the real world, such as manipulating objects and exploring their properties. In general, a haptic interface receives motor commands from the user and displays the appropriate haptic image back to the user. Haptic interactions may be augmented with other forms of stimuli such as stimulation of visual or auditory senses.

Although haptic devices are typically designed for interaction with the hand, there are a number of alternative options that are appropriate for sensory and motor properties of other parts of the body. In general, a haptic interface is a device that: (1) measures position or contact force (and/or their time derivatives and spatial distribution) and (2) displays contact force or position (and/or their spatial and time distribution) to the user.

Figure 2.1 shows a block diagram of a typical haptic system. A human operator is included in the haptic loop through a haptic interface. The operator interacts with a haptic interface either through force or movement. The interface measures human activity. The measured value serves as a reference input either to a teleoperation system or a virtual environment. A teleoperation system is a system in which a usually remote slave robot accomplishes tasks in the real environment that the human operator specifies using the haptic interface. Interaction with a virtual environment is similar, except that both the slave system and the objects manipulated by it are part of the programmed virtual environment. Irrespective of whether the environment is real or virtual, control of the slave device is based on a closed loop system that compares the output of the haptic interface to the measured performance of the slave system. The essence of haptic interaction is the display of forces or movements, which are the result of the operation of the slave system, back to the user through the haptic interface. Therefore, it is necessary to measure forces and movements that occur in teleoperation or compute forces and movements that are the result of interaction with a virtual environment. Since force may be a result of movement dynamics or interactions of an object with other objects or with the slave system, collision detection represents a significant part of the haptic loop. As already mentioned, contact can occur either between objects in the environment (real or virtual) or between an object and the slave system. Collision detection in a real environment is relatively straightforward and is essentially not much more than the measurement of interaction forces between the robot and its surroundings. In contrast, collision detection in a virtual environment is a more complex task since it requires computation of contact between virtual objects that can be modeled using different methods. In this case, it is necessary to compute multiple contacts between outside surfaces of objects.

Collision detection forms the basis for computation of reaction forces. In a teleoperation system, force is measured directly using a force/torque sensor mounted on the slave robot end-effector. In a virtual environment, on the other hand, it is necessary to compute the contact force based on a physical model of the object. The object stiffness can, for example, be modeled as a spring-damper system, while friction can

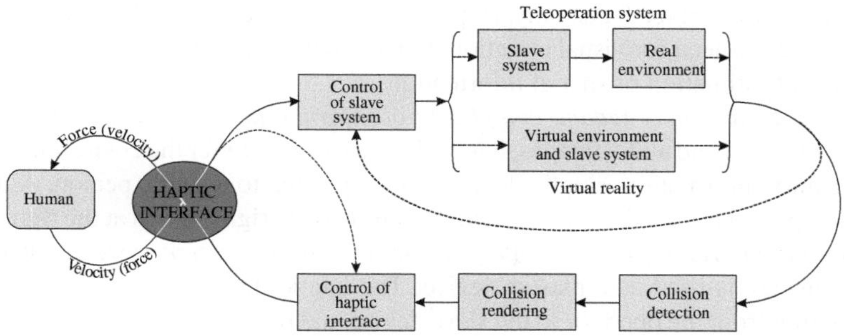

Fig. 2.1 Haptic system: interaction between a human and the haptic interface represents a bidirectional exchange of information—a human operator controls the movement of a slave system as well as receives information about the forces and movements of the slave system through the haptic interface

be modeled as a force that is tangential to the surface of the object and proportional to the normal force to the surface of the object. The computed or measured force or displacement is then transmitted to the user through the haptic interface. A local feedback loop controls the movement of the haptic interface, so that it corresponds to the measured or computed value.

From the block scheme in Fig. 2.1, it is clear that the interaction between a human and the haptic interface represents a bidirectional exchange of information—a human operator controls the movement of a slave system as well as receives information about the forces and movements of the slave system through the haptic interface. The product of force and displacement represents mechanical work accomplished during the haptic interaction. Bidirectional transfer of information is the most characteristic feature of haptic interfaces compared to display of audio and visual images.

2.2 Haptic Applications

The need for an active haptic interface depends on task requirements. *Active haptic interfaces are a must for certain tasks*. A lot of assembly and medical problems are haptic by their nature. Haptic devices are required for simulating such tasks for training purposes, since perception of force, which is the result of the interaction of a tool with the environment, is critical for successful task completion. In addition, haptic devices allow persons with vision impairments to interact with virtual environments.

Haptic devices can improve user immersion. Simple haptic devices with fewer active degrees of freedom are produced in large quantities for entertainment purposes (playing video games). Although the complexity of stimuli that may be transmitted to the user is limited, perception of the virtual environment is still relatively precise.

Haptic devices can improve the efficiency of task execution by providing natural constraints (virtual fixtures). In virtual environments, transfer of virtual objects without haptic perceptions is often difficult. Without feedback information about contact forces, simulation of an assembly task requires a great deal of attention due to reliance on visual feedback only. Haptic devices represent a suitable solution since they reduce the need for visual attention. Force feedback substantially contributes to accuracy of estimation of spatial information.

Haptic devices may reduce complexity of information exchange. In contrast to display of visual and audio images, haptic devices do not clutter the environment with unnecessary information. Haptic devices are connected to a single person. A haptic interface provides only the necessary information to the right person at the right time.

A haptic interface forms an integral part of a teleoperation system, where the haptic display is used as a master device. The haptic interface conveys command information from the operator to the slave device and provides feedback information about the interaction between the slave manipulator and the environment back to the operator.

2.3 Terminology

The terminology is defined as in [4].

A haptic display is a mechanical device designed for transfer of kinesthetic or tactile stimuli to the user. Haptic displays differ in their kinematic structure, workspace and output force. In general, they can be divided into devices that *measure movement and display force* and devices that *measure force and display movement*. The former are called *impedance displays*, while the latter are called *admittance displays*. Impedance displays typically have small inertia and are backdrivable. Admittance displays typically have much higher inertia, are not backdrivable and are equipped with a force and torque sensor.

A haptic interface comprises everything between the human and the virtual environment. A haptic interface always includes a haptic display, control software and power electronics. It may also include a virtual coupling that connects the haptic display to the virtual environment. The haptic interface enables exchange of energy between the user and the virtual environment and it is, therefore, important in the analysis of stability as well as efficiency.

A virtual environment is a computer generated model of a real environment. A virtual environment can be constructed as an exact replica of the real environment or can be a highly simplified reality. Regardless of its complexity, however, there are two completely different ways of interaction between the environment and the haptic interface. Environment may behave as impedance, where the input is the velocity or position and the output force is determined based on a physical model, or as an admittance, where the input is force and the output is velocity or position.

A haptic simulation is a synthesis of a user, haptic interface and a virtual environment. All these elements are important for stability of the system. Simulation includes continuous time elements, such as a human and a mechanical device, as well as discrete elements, such as a virtual environment and control software.

Mechanical impedance is an analogy to electrical impedance. It is defined as the ratio between force and velocity (torque and angular velocity)—an analogy of the ratio between voltage and current in electrical circuits:

$$Z(s) = \frac{F}{v} = ms + b + \frac{k}{s}, \tag{2.1}$$

where m is mass, b is a viscous damping and k is stiffness. Mechanical impedance is often defined as the ratio between force and position (displacement). This definition is related to the second-order differential equation that describes the mechanical system as

$$F = m\ddot{x} + b\dot{x} + kx. \tag{2.2}$$

In this case, impedance is defined as

$$Z(s) = \frac{F}{x} = ms^2 + bs + k. \tag{2.3}$$

Mechanical admittance represents an analogy to electrical admittance and is defined as the ratio of the velocity and force (angular velocity and torque)—an analogy of the ratio between current and voltage:

$$Y(s) = \frac{v}{F} = \frac{1}{ms + b + \frac{k}{s}}, \tag{2.4}$$

where m is the mass, b is the viscous damping and k is the stiffness. Similarly to mechanical impedance, admittance is also often defined as the ratio of position (displacement) and force

$$Y(s) = \frac{x}{F} = \frac{1}{ms^2 + bs + k}. \tag{2.5}$$

Causal structure is defined by the combination of the type of haptic display (impedance or admittance) and the virtual environment (impedance or admittance), giving a total of four possible combinations.

References

1. Minsky, M., Ouh-Young, M., Steele, O., Jr., F.B., Behensky, M.: Feeling and seeing: issues in force display. Computer Graphics, vol. 24, pp. 235–443. ACM Press, New York (1990)
2. Barfield, W., Furness, T.A.: Virtual Environments and Advanced Interface Design. Oxford University Press, New York (1995)
3. Duke, D., Puerta, A.: Design, Specifications and Verification of Interactive Systems. Springer, Wien (1999)
4. Addams, R.J., Hannaford, B.: Stable haptic interaction with virtual environments. IEEE Trans. Robot. Autom. **15**, 465–474 (1999)

Chapter 3
Human Haptic System

A human haptic system can be in general divided into three main subsystems:

- *sensory capabilities*: kinesthetic and tactile senses enable gathering information about the environment through touch;
- *motor capabilities*: the musculoskeletal system allows positioning of the human sensory system for obtaining information about objects and for manipulation of objects through interaction;
- *cognitive capabilities*: the central nervous system analyzes gathered information about the environment and maps it into motor functions based on objectives of the task.

When designing haptic interfaces with the aim of providing an optimal interaction with the human user, it is necessary to understand the *roles* of the motor, sensory and cognitive subsystems of the human haptic system. The mechanical structure of the human hand, for example, consists of a complex arrangement of bones connected by joints and covered with layers of soft tissue and skin. Muscles, which control 22 degrees of freedom of the hand, are connected through tendons to the bones. The sensory system of the hand includes a variety of receptors in the skin, joints, tendons and muscles. Mechanical, thermal or chemical stimuli activate appropriate receptors, triggering the nerve stimuli, which are converted to electrical impulses and relayed by afferent nerves to the central nervous system. From the central nervous system signals are conveyed in the opposite direction by the efferent nervous system to muscles, which execute the desired movement.

In the real world, whenever we touch an object, external forces are generated, which act on the skin. Haptic sensory information, conveyed from hands to the brain during contact with the object, can be divided into two classes: (1) *Tactile information* refers to the perception of the nature of contact with the object and is mediated by low-threshold mechanoreceptors in the skin (e.g. in the finger tip) within and around the contact area. It enables estimation of spatial and temporal variations of the distribution of forces within the area of contact. Fine texture, small objects, softness, slipperiness of the surface and temperature are all perceived by tactile sensors. (2) *Kinesthetic information* refers to the perception of position and movement of a

M. Mihelj and J. Podobnik, *Haptics for Virtual Reality and Teleoperation*, Intelligent Systems, Control and Automation: Science and Engineering 64,

DOI: 10.1007/978-94-007-5718-9_3, © Springer Science+Business Media Dordrecht 2012

limb together with the forces acting on that limb. This perception is mediated by the sensory nerve signals from the receptors in the skin around the joints, in the joints, tendons and muscles. This information is further augmented with motor control signals. Whenever arm movement is employed for environment exploration, kinesthetic information enables perception of natural properties of objects such as shape and compliance or stiffness. Information transmitted during a passive and stationary contact of the hand with an object is predominantly tactile information (kinesthetic information provides details about the position of the arm). On the other hand, during active arm motion in free space (the skin of the hand or of the arm is not in contact with surrounding objects) only kinesthetic information is conveyed (absence of tactile information indicates that the arm is moving freely). In general both types of feedback information are simultaneously present during actively performed manipulation tasks.

When actively performing the task, supervision of contact conditions is as important as the perception of touch. Such supervision involves both fast muscle or spinal reflexes and relatively slow voluntary responses. In motor tasks, such as pinch grasp, motor activity for increasing grasp force occurs in as little as 70 ms after the object slip is detected by finger tips. Human skills for grasping and manipulation are the result of mechanical properties of skin and subcutaneous tissue, which provides rich sensory information from diverse and numerous receptors that monitor the execution of tasks and the ability of nervous system to fuse this information with the activity of the motor system.

The human haptic system is composed of, in addition to tactile and kinesthetic sensory subsystems, a motor system, which enables active exploration of the environment and manipulation of objects, and a cognitive system, which associates action with perception. In general, contact perception is composed of both tactile and kinesthetic sensory information and contact image is constructed by guiding the sensory system through the environment using motor commands that depend on the objectives of the user. Given the large number of degrees of freedom, multiplicity of subsystems, spatial distribution of receptors and sensory-motor nature of haptic tasks, the human haptic capabilities and limitations that determine the characteristics of haptic devices are difficult to determine and characterize.

Haptic devices receive motor commands from the user and display the image of force distribution to the user. A haptic interface should provide a good match between the human haptic system and hardware equipment used for sensing and displaying haptic information. The primary input–output (measured and displayed) variables of the haptic interface are movement and force (or vice versa), with their spatial and temporal distributions. Haptic devices can therefore be treated as generators of mechanical impedance, which represents the relation between the force and movement (and their derivatives) in various positions and orientations. When displaying contact with a finite impedance, either force or movement represent excitation, while the remaining quantity represents the response (if force is excitation then movement is response and vice versa), which depends on the implemented control algorithm. Consistency between the free movement of hands and touch is best achieved by taking into account the position and movement of hands as excitation and resultant vector of force and its distribution within the area of contact as response.

Since a human user senses and controls the position and force displayed by a haptic device, the performance specifications of the device directly depend on human capabilities. In many simple tasks that involve active touch, either tactile or kinesthetic information is of primary importance, while the other is only complementary information. For example, when trying to determine the length of a rigid object by holding it between thumb and index finger, essential information is kinesthetic information, while tactile information is only supplementary information. In this case the crucial ability is sensing and controlling the position of the finger. On the other hand, perception of texture or slipperiness of the surface depends mainly on tactile information, while kinesthetic information only supplements tactile perception. In this case, perceived information about temporal-spatial distribution of forces provides a basis for perceiving and inferring about the conditions of contact and characteristics of the surface of the object. In more complex haptic tasks, however, both kinesthetic and tactile feedback is required for correct perception of the environment.

Due to hardware limitations, haptic interfaces can provide stimuli that only approximate interaction with a real environment. However, this does not mean that an artificially synthesized haptic stimulus does not feel realistic. Consider the analogy with a visual experience of watching a movie. Although visual stimuli in the real world are continuous in time and space, visual displays project images with a frequency of only about 30 frames per second. Nevertheless, the sequence of images is perceived as a continuous scene, since displays are able to exploit limitations of the human visual apparatus.

Similar reasoning applies also for haptic interfaces, where implementation of appropriate simplifications that are relevant for the given task, exploits limitations of the human haptic system. Understanding of human biomechanical, sensory-motor and cognitive capabilities is critical for proper design of device hardware and control algorithms for haptic interfaces.

Compared to vision and hearing, our understanding of human haptic system, which includes both sensory and motor systems is very limited. One of the reasons is the problem of empirical analysis of the haptic system owing to difficulties in delivering appropriate stimuli because of the bidirectional nature of the human haptic system. Fundamental issues in the analysis of human sensory system are: (1) perception of forces in quasi-static and dynamic conditions, (2) perception of pressure, (3) position sensing resolution and (4) level of stiffness required for a realistic display of a rigid environment. Fundamental issues in the analysis of motor system performance are: (1) the maximum force that a human can produce with different body segments, (2) accuracy of control of the force applied by the human on the environment and (3) force control bandwidth. Moreover, important issues are also ergonomics and comfort of haptic devices.

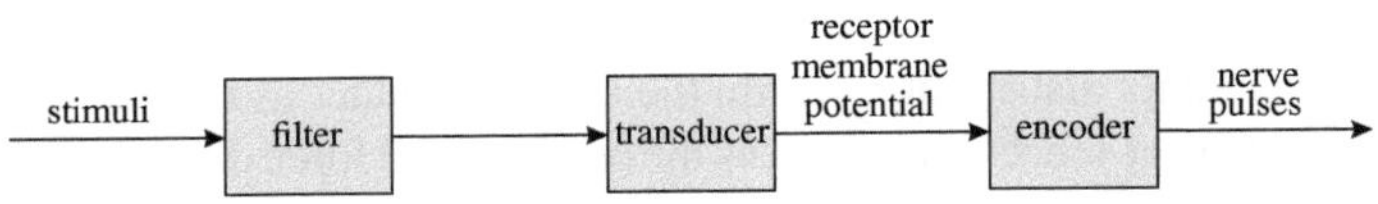

Fig. 3.1 Receptor model block diagram

3.1 Receptors

Biological transducers that respond to stimuli coming from the environment or from within the human body and transmit signals to the central nervous system are known as receptors. There are different types of receptors in the human body and each receptor is generally sensitive only to one type of energy or stimulus. The structure of receptors is therefore very heterogeneous and each receptor is adapted to the nature of stimuli that trigger its response. Despite the diversity of receptor morphology, the structure of the majority of receptors can be divided into three functional subsystems shown in Fig. 3.1.

Input signal is a stimulus, which occurs in one of the following forms of energy: electromagnetic, mechanical, chemical or thermal. A stimulus acts on the filter part of the receptor, which does not change the form of energy, but amplifies or attenuates some of the stimuli parameters. For example, the shape of the outer ear amplifies certain frequencies, the skin acts as a mechanical filter, and the lens in the eye focuses light rays on the retina. A transducer changes filtered stimuli into receptor membrane potential and an encoder encodes the amplitude of the membrane potential into a sequence of nerve pulses.

In general, receptor output decreases toward background level when a constant stimulus is present for an extended time. This phenomenon is called adaptation. Receptor response can in general be decomposed into two components: the first component is proportional to stimulus intensity and the second is proportional to the rate of change of the stimulus. For receptor response $R(t)$ and for stimulus intensity $S(t)$ the following relation applies:

$$R(t) = \alpha S(t) + \beta \dot{S}(t), \tag{3.1}$$

where α and β are constants or functions describing the adaptation.

For many types of stimuli it has been observed that responses can be described by the *Weber-Fechner* equation

$$R = K \log \frac{S}{S_0}, \tag{3.2}$$

where K is a constant and S_0 is a threshold value of the stimulus.

Human senses are divided into two main categories: somatosensory senses and special senses. Special senses include vision, hearing, smell, taste and balance and will not be addressed here. Somatosensory senses collect information from the stimuli acting on the surface of the body or originating within the body. Somatosensory senses are divided into mechanoreceptors, thermoreceptors and nociceptors (receptors for pain). The most important receptors for understanding and analysis of haptic interaction are mechanoreceptors, which include receptors for touch, pressure, vibration as well as position and velocity of body segments.

3.2 Kinesthetic Perception

The term kinesthetics refers to the perception of movement and position of limbs and in a broader sense includes also perception of force. This perception originates primarily from mechanoreceptors in muscles, which provide the central nervous system with information about static muscle length, muscle contraction velocity and forces generated by muscles. Awareness of limb position in space, of limb movement and of mechanical properties (such as mass and stiffness) of objects with which the user interacts, emerges from these signals. Sensory information about the change of limb position originates also from other senses, particularly from receptors in joints and skin. These senses are particularly important for kinesthetics of the arm. Receptors in the skin significantly contribute to the interpretation of the position and movement of the arm. The importance of cutaneous sensory information is not surprising considering the high density of mechanoreceptors in the skin and their specialization for tactile exploration. This feedback information is important for kinesthetics of the arm because of the complex anatomical layout of muscles that extend across a number of joints, which introduces uncertainty in the perception of position derived from receptors in muscles and tendons.

3.2.1 Kinesthetic Receptors

Mechanoreceptors, which are found in muscles, are primary and secondary receptors (also called Type Ia and Type II sensory fiber) located in muscle spindles. Muscle spindles are elongated structures 0.5–10 mm in length, made up of bundles of muscle fibers. Spindles are located parallel to the muscle fibers, which are generators of muscle force and are attached at both ends either to the muscle or tendon fibers [1]. A muscle spindle detects length and tension changes in muscle fibers. The main role of a muscle spindle is to respond to stretching of the muscle and to stimulate muscle contraction through a reflex arc to prevent further extension. Reflexes play an important role in the control of movement and balance. They allow automatic and rapid adaptation of muscles to changes in load and length.

Both primary and secondary spindle receptors respond to changes in muscle length. However, the primary receptors are much more sensitive to velocity and acceleration components of the movement and their response considerably increases with increased velocity of muscle stretching. The response of primary spindle receptors

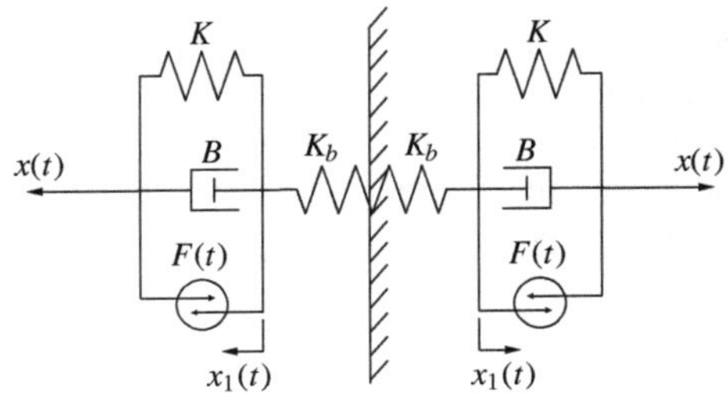

Fig. 3.2 Biomechanical model of the muscle spindle

is nonlinear and their output signal depends on the length of the muscle, muscle contraction history, current velocity of muscle contraction and activity of the central nervous system, which modifies the sensitivity of muscle spindles. Secondary spindle receptors have a much less dynamic response and have a more constant output at constant muscle length compared to the primary receptors. Higher dynamic sensitivity of primary spindle receptors indicates that these receptors mainly respond to the velocity and direction of muscle stretching or movement of a limb, while the secondary spindle receptors measure static muscle length or position of the limb.

A biomechanical model of a muscle spindle is shown in Fig. 3.2 [2]. The model consists of a series of elastic element K_b, which represent predominantly the elastic central part of the nucleus follicle, and parallel connection of elastic element K, viscous element B and active element F, which generates force. Suppose that both ends of the spindle are stretched for x. Since both ends are equally stretched, the center of the nucleus follicle does not move. Therefore, only half of the muscle spindle can be considered for mathematical model derivation. Thus, we can write the following equation

$$F(t) + B(\dot{x}(t) - \dot{x}_1(t)) + K(x(t) - x_1(t)) = K_b x_1(t). \tag{3.3}$$

Signal from the spindle is proportional to the stretch of the nucleus follicle x_1. Laplace transformation of (3.3) yields

$$\begin{aligned} &F(s) + sBx(s) - sBx_1(s) + Kx(s) - Kx_1(s) = K_b x_1(s) \\ &(K_b + K + sB)x_1(s) = F(s) + (K + sB)x(s) \\ &x_1(s) = \frac{K + sB}{K_b + K + sB}x(s) + \frac{1}{K_b + K + sB}F(s). \end{aligned} \tag{3.4}$$

From the last equation it can be seen that the signal from muscle spindle consists of two components. The contribution due to extension of the muscle depends on the velocity and amount of stretch and the contribution due to innervation depends on the force F.

A higher density of mechanoreceptors is associated with a better resolution of the tactile system. However, this does not apply to the kinesthetic system where the total number of receptors is much lower and a higher density of receptors is not necessarily

Fig. 3.3 Biomechanical model of a muscle with the Golgi tendon organ

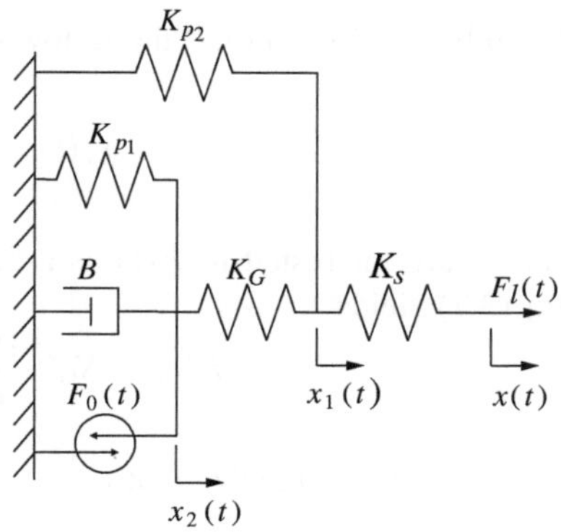

associated with better kinesthetic capabilities. The number of muscle spindles in the muscle depends more on the size of the muscle than on the function of the muscle.

The second type of mechanoreceptors is a Golgi tendon organ. It measures 1 mm in length, has a diameter of 0.1 mm and is located at the attachment of a tendon to the bundle of muscle fibers. The receptor is therefore connected in series with the group of muscle fibers and it primarily responds to the force generated by these fibers. When muscle is exposed to excessive load, the Golgi tendon organ becomes excited, which leads to the inhibition of motor neurons and finally to reduction of muscle tension. In this way, the Golgi tendon organ also serves as a safety mechanism that prevents damage to the muscles and tendons due to excessive loads.

The biomechanical model of a muscle with a Golgi tendon organ receptor is shown in Fig. 3.3 [2]. The model consists of a parallel connection of a muscle elasticity K_p, a muscle viscosity B and of an active element $F_0(t)$ that generates force. In series to the model of the muscle are connected the elasticity of the Golgi tendon organ K_G and elasticity of muscle and tendon K_s. Parallel elasticity of a muscle is split into two components, K_{p_1} and K_{p_2}, where the latter bypasses the Golgi tendon organ and is linked directly to the tendon.

Let us now examine the effect of changes in $x(t)$, which is the total length of the muscle with a tendon, and changes in $x_2(t)$, which is the length of the muscle fibers, on the tension $F(t)$ in the Golgi tendon organ. For static conditions, taking into consideration the elasticity of the Golgi tendon organ K_G, the following equations can be derived:

$$K_s(x(t) - x_1(t)) = K_{p_2}x_1(t) + K_G(x_1(t) - x_2(t)) \tag{3.5}$$

and

$$K_G(x_1(t) - x_2(t)) = K_{p_1}x_2(t) + F_0. \tag{3.6}$$

The tendon tension $F(t)$ is given by

$$F(t) = (x_1(t) - x_2(t))K_G. \tag{3.7}$$

From Eqs. (3.5) and (3.7) the following relation can be computed:

$$x_1(t) = \frac{K_s x(t) - F(t)}{K_s + K_{p_2}}. \tag{3.8}$$

As we are interested in $F(t)$ as a function of $x(t)$ and $x_2(t)$, inserting $x_1(t)$ into Eq. (3.7) yields

$$F(t) = K_G \frac{K_s x(t) - F(t)}{K_s + K_{p_2}} - K_G x_2(t), \tag{3.9}$$

which can be reorganized into

$$F(t) = \frac{K_s K_G}{K_s + K_G + K_{p_2}} x(t) - \frac{K_G(K_s + K_{p_2})}{K_s + K_G + K_{p_2}} x_2(t). \tag{3.10}$$

Derivation of $F(t)$ as a function of $x(t)$ and $x_2(t)$ does not depend on K_{p_1} and $F_0(t)$, because these two variables determine $x_2(t)$. The Golgi tendon organ has no particular dynamic properties. Its response is proportional to force $F(t)$.

Other mechanoreceptors found in joints are Ruffini endings, which are responsible for sensing angle and angular velocity of the joint movements, Pacinian corpuscles, which are responsible for estimation of the joint acceleration, and free nerve endings, which constitute the nociceptive system of the joint.

3.2.2 Perception of Movements and Position of Limbs

Haptic interaction is based on three basic types of perceptions: perception of limb position, perception of motion and perception of force. Motion perception capabilities depend on various factors, such as velocity of movement, particular joints involved in the movement and a level of contraction of muscles that are involved in movement of the particular joint [1, 3]. Humans can perceive rotation on the level of a fraction of a degree in time frame of 1 s. It is easier to detect fast movements than slower movements (for finger movements, perception threshold drops from 8 to 1°, when the velocity changes from 1.25 to 10°/s). It is easier to detect movements in proximal joints such as elbow and shoulder joints than movements of the same magnitude in more distal joints [4]. The minimum detectable change is approximately 2.5° for finger joints, 2° for the wrist and elbow and 0.8° for the shoulder joint. Better capabilities in detection of small changes in proximal joint are not coincidental, because proximal joints tend to move slower than the distal joints and the same joint angle error in proximal joints results in higher positional error at the tip of the limb. For example, 1° rotation of the shoulder joint with a fully extended arm results in the displacement of the tip of the middle finger of 13 mm, while 1° rotation of the distal joint of the middle finger results in the displacement of the tip of the middle finger of 0.5 mm. In contrast to sensing motion of the limbs the capability to detect change in position of the limb depends only on the absolute position of the limb and is independent of the velocity of the movement (Table 3.1).

3.2.3 *Perception of Force*

The contact force when touching an object is perceived through tactile and kinesthetic sensory modalities. The outputs of the Golgi tendon organ provide information about force exerted by muscles. The smallest change of force perceived by a human is a function of currently applied force [4]. The differential threshold of perceived change of force ranges from 5 to 12 % for a range of forces between 0.5 and 200 N and is constant for a variety of muscle groups. The differential threshold increases to the range of 17–27 % for forces lower than 0.5 N (Table 3.1).

3.2.4 *Perception of Stiffness, Viscosity and Inertia*

A kinesthetic system is involved not only in the acquisition of information related to the forces generated by the muscles and a resulting movement of the limbs, but also uses this information to evaluate quantities such as stiffness, viscosity and inertia, for which humans possess no specific sensors. Sensing of these quantities is of particular importance for the design of haptic devices, since their mechanical properties have a significant impact on the efficiency of the human operator.

Resolution of stiffness and viscosity perception is relatively poor compared to the resolution of force and displacement perception (both quantities are required for stiffness estimation) or to the resolution of force and velocity of movement perception (both quantities are required for viscosity estimation) [1]. The differential threshold for detecting a change of stiffness is 8–22 % (to perceive an object as rigid, stiffness of at least 25 N/mm is required). The differential threshold for detection of change of viscosity is 14–34 %. Resolution of estimated inertial properties is also relatively poor. The differential threshold for detecting changes of mass is approximately 21 %, while the differential threshold for detecting changes of the object inertia is between 21 and 113 %. The latter value depends on the nominal inertia, which is used to measure the threshold (Table 3.1).

3.3 Tactile Perception

Although humans are presented with various sensations when touching objects, these sensations are a combination of only a few basic types of sensations, which can be represented with basic building blocks. Roughness, lateral skin stretch, relative tangential movement and vibrations are the basic building blocks of sensations when

Table 3.1 Perceptual properties of the kinesthetic senses [1]

Quantity	Resolution	[Differential] threshold
Limb movement	0.5–1° (at 10–80°/s)	8 % (range: 4–19 %)
Limb position	0.8–7°	7 % (range: 5–9 %)
Force	0.6 N	7 % (range: 5–12 %)
Stiffness	–	17 % (range: 8–22 %)
Viscosity	–	19 % (range: 14–34 %)
Inertia	–	28 % (range: 21–113 %)

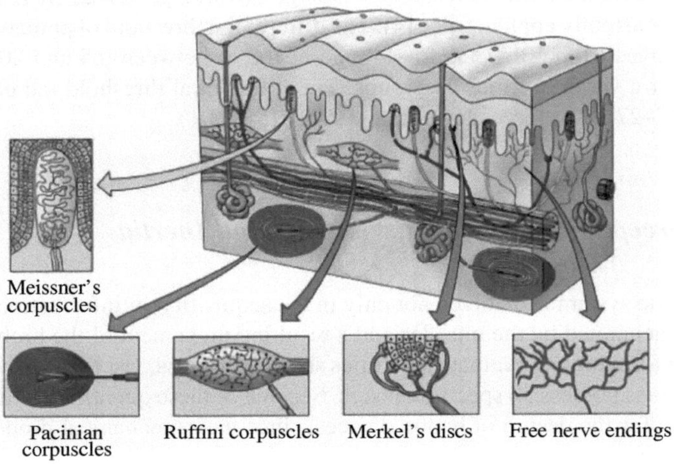

Fig. 3.4 Mechanoreceptors in the skin

touching objects. Texture, shape, compliance and temperature are the basic object properties that are perceived by touch. Perception is based on mechanoreceptors in the skin. When designing a haptic device, human temporal and spatial sensory capabilities have to be considered.

Four different types of sensory organs for sensing touch can be found in the skin. These are Meissner's corpuscles, Pacinian corpuscles, Merkel's discs and Ruffini corpuscles (Fig. 3.4). Figure 3.5 shows the rate of adaptation of these receptors to stimuli, the average size of the sensory area, spatial resolution, sensing frequency range and frequency of maximum sensitivity. Delay in the response of these receptors ranges from 50 to 500 ms.

The thresholds for different receptors overlap and hence, the quality of sensing of touch is determined by a combination of responses of different receptors. Receptors complement each other, making it possible to achieve a wide sensing range for detecting vibrations with frequencies ranging from 0.4 to about 1000 Hz [3, 5]. In general, the threshold for detecting tactile inputs decreases with increased duration of the stimuli. The spatial resolution at the fingertips is about 0.15 mm, while the

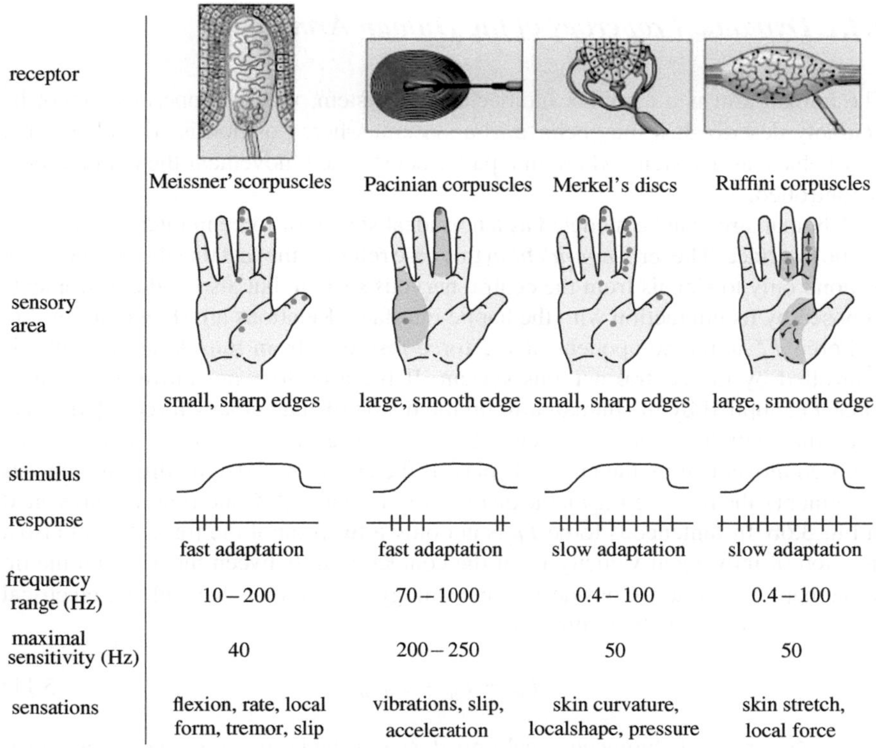

Fig. 3.5 Functional properties of skin mechanoreceptors

minimum distance between two points that can be perceived as separate points is approximately 1 mm. Humans can detect a 2 μm high needle on the smooth glass surface. Skin temperature affects the tactile perception.

Properties of the human tactile perception provide important guidelines for planning and evaluation of tactile displays. Size of perception area, duration and frequency of the stimulus signal needs to be considered.

3.4 Human Motor System

During haptic interactions a user directly interacts with a haptic display through physical contact. In consequence this affects the stability of haptic interaction. It is therefore necessary to consider human motor properties to ensure stable haptic interaction.

3.4.1 Dynamic Properties of the Human Arm

The human arm is a complex biomechanical system, whose properties cannot be uniquely described; it may behave as a system, where position is controlled, or it may behave as a system, where in a partly constrained movement the contact force is controlled.

A human arm can be modeled as a non-ideal source of force in interaction with a haptic interface. The term *non-ideal* in this case refers to the fact that the arm does not respond only to signals from the central nervous system, but also to the movements imposed by its interaction with the haptic interface. Relations are shown in Fig. 3.6.

Force $F_h^\star$ is the component of the force resulting from muscle activity that is controlled by the central nervous system. If the arm does not move, the contact force F_h, applied by the human arm on the haptic display equals force $F_h^\star$ (muscle force that initializes the movement). However, force F_h is also a function of the imposed movement by the haptic display. If the arm moves (haptic display imposes movement), the force acting on the display differs from $F_h^\star$. Conditions are presented in Fig. 3.6b. Instantaneous force F_h is not only a function of the force $F_h^\star$ but also a function of movement velocity v_h of the contact point between the arm and the tip of the haptic interface. Considering the analogy between mechanical and electrical systems, force F_h can be written as

$$F_h = F_h^\star - Z_h v_h, \tag{3.11}$$

where Z_h represents biomechanical impedance of the human arm and maps movement of the arm into force. Z_h is primarily determined by physical and neurological properties of the human arm and has an important role in the stability and performance of the haptic system.

Dynamics of arm force generation is governed by the following subsystems (Fig. 3.6a) [6]:

1. G_a represents the dynamics of muscle activation, which generates muscle force $F_h^\star$ as a response to the commands of the central nervous system u,
2. G_p (dynamics of muscular contraction and passive tissue) represents properties of muscle and passive tissue, which surrounds joints and reduces muscle force $F_h^\star$ for $G_p v_h$,
3. G_f represents dynamics of the neural feedback loop, which controls the forces applied by the human arm on the haptic interface.

3.4.2 Dynamics of Muscle Activation

Transfer function G_a maps central nervous system commands u into muscular force $F_h^\star$. A simplified dynamics of muscle activation G_a can be represented with a first-order transfer function with time dependent time constant, which depends on the input u.

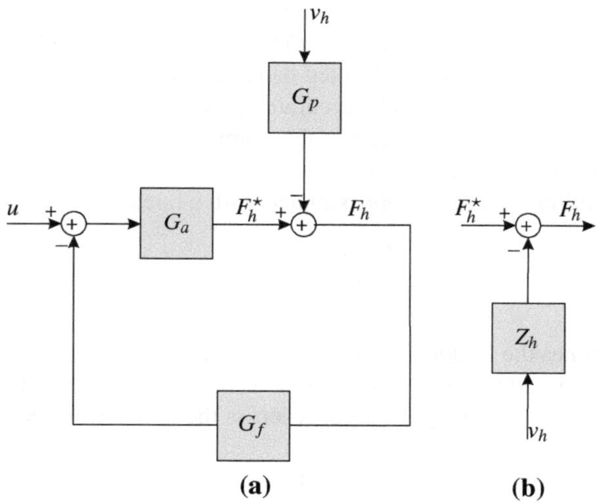

Fig. 3.6 **a** Internal structure of the human arm impedance, **b** contact force F_h is a function of muscular force $F_h^\star$ and arm impedance Z_h

3.4.3 Dynamics of Muscle Contraction and Passive Tissue

Transfer function G_p determines the biomechanical impedance of the human arm. The function implicitly takes into account the muscle internal dynamics and variable stiffness of the arm due to simultaneous contraction of antagonistic muscles. G_p also includes the dynamics of passive tissue surrounding the joints. Equation (3.12) represents a general form of the transfer function G_p,

$$G_p = m_a s^2 + b_a s + k_a, \tag{3.12}$$

where m_a represents mass of the arm, while b_a and k_a represent viscous and elastic properties of muscles and passive tissue.

It should be noted that the signals from the central nervous system to the arm have a dual role: (1) they determine the movement trajectory of the arm and (2) they change the biomechanical impedance G_p of the arm. The first function is denoted as u in Fig. 3.6a, while the second function, which changes the impedance of the arm, is not explicitly shown. It is implicitly included in the G_p.

3.4.4 Neural Feedback Loop

Transfer function G_f represents the neural feedback loop, which enables fine control of the force applied by the user on the haptic interface. The interaction force F_h

applied by the arm on the haptic interface is used as a signal, which modulates the commands from the central nervous system to the arm. A feedback loop is effective only at low frequencies due to the limited bandwidth of the central nervous system. The user is able to perform highly accurate force interaction at low frequencies, however as movements become faster, the control of contact force becomes less accurate and therefore the gain of the transfer function G_f decreases with increasing frequency. Function G_f can be approximated with a linear first-order relation

$$G_f = \frac{C}{s + \lambda}, \tag{3.13}$$

where λ determines the frequency bandwidth of the central nervous system and C is a feedback gain. In addition, a delay can be inserted into the transfer function that represents delays in conveying the signals across the nervous system

$$G_f = \frac{Ce^{-sT}}{s + \lambda}, \tag{3.14}$$

where T represents the time delay.

The block diagram in Fig. 3.6a can be simplified into diagram 3.6b and written in equation form as

$$F_h = F_h^\star - Z_h v_h. \tag{3.15}$$

By taking into account $F_h^\star = G_{CNS} u$, the resulting force equals

$$F_h = G_{CNS} u - Z_h v_h, \tag{3.16}$$

where

$$G_{CNS} = \frac{G_a}{1 + G_f G_a} \tag{3.17}$$

and

$$Z_h = \frac{G_p}{1 + G_f G_a}. \tag{3.18}$$

The transfer function G_{CNS} represents the effects of central nervous system commands and Z_h represents the effect of the arm movement in the contact point with the haptic interface. From the expression (3.18) it is evident that the transfer function Z_h is not merely a biomechanical impedance, but also properties of the nervous system should be considered. Force applied by the human arm on the device is a result of both the central nervous system commands and the movements of the device.

The bandwidth of the human motor system depends on the type of activity: 1–2 Hz for responses to unexpected stimuli; 2–5 Hz for periodic movements; up to 5 Hz for the generated or learned trajectories; and up to 10 Hz for the reflex responses.

3.5 Special Properties of the Human Haptic System

The human haptic system possesses certain properties that are relevant for the design of haptic interfaces [3].

The gradient in tactile resolution is an important feature of the human sensory system. Using the tip of the finger humans can detect fine surface texture, high frequency vibrations, they can distinguish between two closely separated points and detect fine translational motion. At more proximal locations on the limbs and the trunk the sensitivity is considerably lower. The principle of gradient from distal to proximal applies as a general rule, with exceptions of the lips and the tongue.

The gradient in detecting movements is an important feature of a human motor system. A similar effect as the gradient in the tactile resolution can also be seen in detecting displacement of the fingertips. If movement is limited to the distal joint of the finger, very small limb endpoint displacements can be detected. However, if movement is limited to the elbow joint or to the shoulder, the displacement must be a few times larger in order to be detected.

In this regard, a clear gradient in capabilities can be seen, as skin closer to the fingertips allows a better sensation and segments closer to the fingertips can be more accurately controlled than those closer to the trunk.

In general, tactile and kinesthetic receptors in the skin and muscles behave as second-order systems, which means that their response depends on the amplitude and the rate of change of stimuli. Therefore, in short time intervals small displacements can be perceived. On the other hand, in longer intervals a human haptic system neither detects nor compensates for large positional errors. This low discrimination in perception of the absolute position allows small position corrections in the haptic devices without attracting the attention of the user or degrading the quality of interaction. In this way, it is possible to compensate for the limited workspace of most haptic devices.

A natural way for humans to perform various tasks is to first set the reference coordinate frame with the non-dominant limb and then to carry out the task with the dominant limb in this reference coordinate frame. This reference frame affects the accuracy and velocity of the user when using a haptic interface. Thus, bimanual interfaces allow improvements in speed and precision.

References

1. Jones, L.A.: Kinesthetic sensing. In: Human and Machine Haptics. MIT Press, Cambridge (2000)
2. Vodovnik, L.: Osnove biokibernetike. Fakulteta za elektrotehniko, Ljubljana (1968)
3. Biggs, S.J., Srinivasan, M.A.: Haptic interfaces. Handbook of Virtual Environments. LA Earlbaum, New York (2002)
4. Tan, H.Z., Srinivasan, M.A., Eberman, B., Cheng, B.: Human factors for the design of force-reflecting haptic interfaces. Dyn. Syst. Control **55**, 353–359 (1994)
5. Ederman, L.S., Klatzky, R.: Haptic perception: a tutorial. Atten. Percept. Psychophys. **71**, 1439–1459 (2009)
6. Kazerooni, H., Her, M.G.: The dynamics and control of a haptic interface device. IEEE Trans. Robot. Autom. **20**, 453–464 (1994)

Chapter 4
Haptic Displays

4.1 Kinesthetic Haptic Displays

Haptic displays are devices composed of mechanical parts, working in physical contact with a human body for the purpose of exchanging information. When executing tasks with a haptic interface the user transmits motor commands by physically manipulating the haptic display, which in the opposite direction displays a haptic sensory image to the user via correct stimulation of tactile and kinesthetic sensory systems. This means that haptic displays have two basic functions: (1) to measure positions and interaction forces (and their time derivatives) of the user limb (and/or other parts of the human body) and (2) display interaction forces and positions (and their spatial and temporal distributions) to the user. The choice of a quantity (position or force) that defines motor activity (excitation) and haptic feedback (response) depends on the hardware and software implementation of the haptic interface as well as on the task for which the haptic interface is used [1–3].

4.1.1 Criteria for Design and Selection of Haptic Displays

A haptic display must satisfy at least a minimal set of kinematic, dynamic and ergonomic requirements in order to guarantee adequate physical efficiency and performance with respect to the interaction with a human operator.

4.1.1.1 Kinematics

A haptic display must be capable of exchanging energy with the user across mechanical quantities, such as *force* and *velocity*. The fact, that both quantities exist simultaneously on the user side as well as on the haptic display side, means that the haptic display mechanism must enable a continuous contact with the user for the whole time, when the contact point between the user and the device moves in a three-dimensional space.

M. Mihelj and J. Podobnik, *Haptics for Virtual Reality and Teleoperation*, Intelligent Systems, Control and Automation: Science and Engineering 64,
DOI: 10.1007/978-94-007-5718-9_4, © Springer Science+Business Media Dordrecht 2012

The most important kinematic parameter of a haptic display is the number of degrees of freedom. In general, the higher the number of degrees of freedom, the greater the number of directions in which it is possible to simultaneously apply or measure forces and velocities. Number of degrees of freedom, the type of degrees of freedom (rotational or translational joints) and the length of the segments determine the workspace of the haptic display. In principle, this should include at least a subset of the workspace of human limbs, but its size primarily depends on the tasks for which it is designed.

An important aspect of kinematics of a haptic display presents the analysis of singularities [4]. The mechanism of the display becomes singular, when one or more joints are located at the limits of their range of motion or when two or more joint axes become collinear. In a singular pose the mechanism loses one or more of its degrees of freedom.

A Jacobian matrix relates joint velocities $\dot{\mathbf{q}}$ with end-effector velocities $\dot{\mathbf{x}}$, $\dot{\mathbf{x}} = \mathbf{J}(\mathbf{q})\dot{\mathbf{q}}$. The Jacobian matrix becomes singular close to the singularity point of the mechanism. Consequently, the inverse Jacobian matrix cannot be computed and relation $\dot{\mathbf{q}} = \mathbf{J}^{-1}(\mathbf{q})\dot{\mathbf{x}}$ does not have a physically meaningful result. Thus, for an arbitrary end-effector velocity $\dot{\mathbf{x}} \neq 0$, one or more joint velocities $\dot{\mathbf{q}}$ approach infinity, when $\mathbf{q}$ approaches singularity.

Analysis based on the use of a Jacobian matrix can be extended to analysis of forces and torques. Recall the relation between joint torques $\boldsymbol{\tau}$ and the force applied at the mechanism end-effector $\mathbf{F}$, $\boldsymbol{\tau} = \mathbf{J}^T(\mathbf{q})\mathbf{F}$. In singular pose, a Jacobian matrix loses its full rank, meaning that it becomes impossible to apply a controlled force or torque in one or more orthogonal directions. The inverse relation $\mathbf{F} = \mathbf{J}^{-T}(\mathbf{q})\boldsymbol{\tau}$ shows that, when approaching singularity, end-effector forces or torques in one or more directions approach infinity.

4.1.1.2 Dynamics

The intrinsic haptic display dynamics distorts forces and velocities that should be displayed to the user. A convincing presentation of contact with a stiff object, for example, requires high frequency response bandwidth of a haptic system. Thus, persuasiveness of force and velocity rendering is limited by the intrinsic dynamics of the haptic display. The effect of the intrinsic dynamics can be analyzed in a case study with a simplified haptic device consisting of a single degree of freedom as shown in Fig. 4.1 [4]. A haptic display applies force on a user, while the user determines the movement velocity. An ideal display would allow undistorted transfer of a desired force ($F = F_a$; F_a is the actuator force and F is the force applied on the user) and precise velocity measurement ($\dot{x}_m = \dot{x}$; $\dot{x}$ is the actual velocity of the system endpoint and $\dot{x}_m$ is the measured velocity of the system endpoint). However, by taking into account the haptic display dynamics, the actual force applied on the user equals

$$F = F_a - F_f(x, \dot{x}) - m\ddot{x}. \tag{4.1}$$

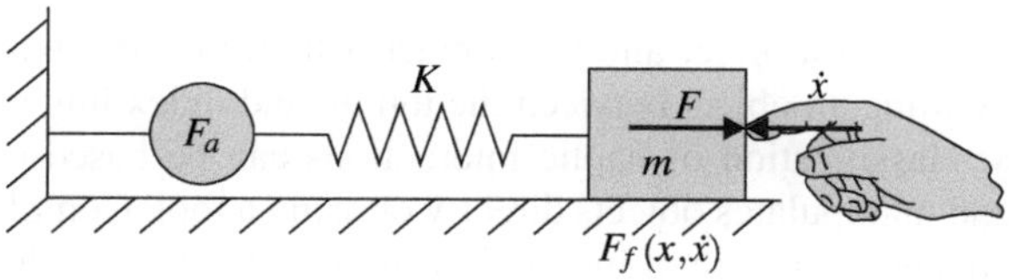

Fig. 4.1 Dynamic model of a haptic display with a single degree of freedom (adapted from [4])

Thus, the force perceived by the user is reduced for the effect of friction $F_f(x, \dot{x})$ and inertia m of the haptic display. In this simplified example the stiffness K does not affect the transfer of forces.

Equation (4.1) indicates that the mass of the haptic display affects the transmission of force to the user by resisting the change of velocity. This opposing force is proportional to the acceleration of the display. Minimization of a haptic display mass is necessary, since during collisions with virtual objects large accelerations (decelerations) can be expected. In case of multidimensional displays the dynamics becomes more complex. Except in specific cases, where the dynamics of the mechanism is uncoupled (Cartesian mechanism), in addition to inertia, also Coriolis and centripetal effects cause absorption of actuation forces at velocities different from zero. A haptic display must be able to support its own weight in the gravitational field, as otherwise, the gravitational force that is not associated with the task is transferred to the user. Gravity compensation can be achieved either actively through actuators of the display or passively with counterweights, which further increase inertia of the display.

Equation (4.1) indicates that part of the forces being generated by the actuators are absorbed due to friction. Friction occurs where two surfaces that are in physical contact move against each other. In general, friction can be decomposed into three components: static friction (a force that is required to initiate the motion between two surfaces, one against the other), Coulomb friction that is velocity independent and viscous damping that is proportional to the velocity.

Stiffness of a haptic display determines how the mechanism deforms under static or dynamic loads. Even though final stiffness does not cause absorption of dynamic actuation forces, low mechanism stiffness can have negative effects, as it prevents accurate measurement of haptic display end-effector velocity. Velocity measurements are generally performed using joint optical encoders, thus segment deformation may result in inaccurate velocity measurements that in certain circumstances may lead to unstable haptic interaction.

4.1.2 Classification of Haptic Displays

Haptic interactions that affect the design of haptic displays can be divided into three categories: (1) free movement in space without physical contact with surrounding objects, (2) contact, which includes unbalanced reaction forces, such as pressing on an object with the tip of a finger and (3) contact, which includes balanced internal forces, such as holding an object between the thumb and index finger [5, 6].

Alternatively, classification of haptic interactions can be based on whether the user perceives and manipulates objects directly or using a tool. Complexity of haptic displays highly depends on type of interactions to be simulated by the interface. An ideal haptic display designed for realistic simulation would have to be capable of simulating the handling of various tools. Such a display would measure limb position and display reaction forces. It would have a unique shape (e.g. exoskeleton) that could be used for different applications by adapting the device controller. However, complexity of human limbs and exceptional sensitivity of skin receptors together with inertia and friction of the device mechanism and constraints related to sensing and actuation of the display, prevent the implementation of such complex device based on the state-of-the-art technology.

Haptic displays can be divided into grounded or non-mobile devices and mobile devices. Haptic perception and manipulation of objects require application of force vectors on the user in different points of contact with an object. Consequently equal and opposite reaction forces act on the haptic display. If these forces are internally balanced, as while grasping an object with the index and thumb fingers, then mechanical grounding of the haptic display against the environment is not required. In the case of internally unbalanced forces, as while touching an object with a single finger, the haptic display must be grounded for balancing the reaction forces. This means that a haptic display placed on a table is considered a grounded device, while an exoskeleton attached to the forearm is a mobile device. If the exoskeleton is used for simulating contact with a virtual object using a single finger, forces that would in a real world be transferred across the entire human musculoskeletal system, are now transferred only to the forearm.

Use of grounded haptic displays has several advantages while executing tasks in a virtual or a remote (teleoperated) environment. Such displays can render forces that originate from grounded sources without distortions and ambiguities. They may be used for displaying geometric properties of objects such as size, shape and texture as well as dynamic properties such as mass, stiffness and friction.

The main advantage of mobile haptic displays is their mobility and therefore, larger workspace. In order to illustrate ambiguities while displaying reaction forces using a mobile haptic display, two examples are analyzed in Fig. 4.2: grasping of a virtual ball and pressing a virtual button. In the case of a virtual ball grasped with the thumb and index fingers, forces acting on the tip of fingers are all that is necessary for a realistic presentation of size, shape and stiffness of a virtual object. Only internally balanced forces act between the fingers and the ball. On the other hand, when pressing against a button, the user does not only feel the forces acting on the finger. The reaction forces also prevent further hand movement in the direction of the button. In this case the ungrounded haptic display can simulate the impression of a contact between the finger and the button, but it cannot generate the reaction force that would stop the arm movement [7].

Figure 4.3 shows classification of haptic displays based on their workspace, power and accuracy.

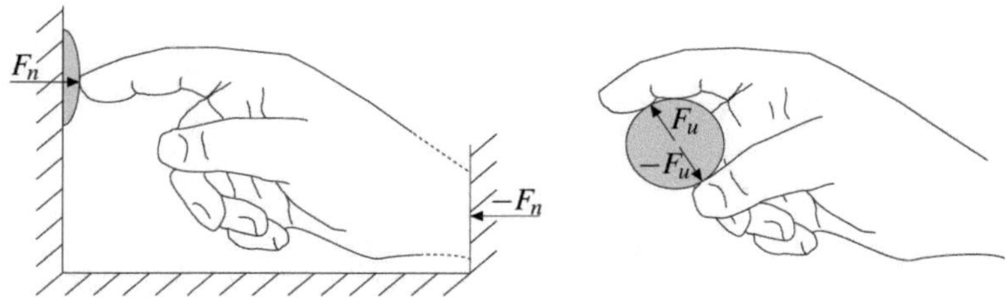

Fig. 4.2 Internally balanced forces F_u when holding a ball and unbalanced forces F_n when pressing a button

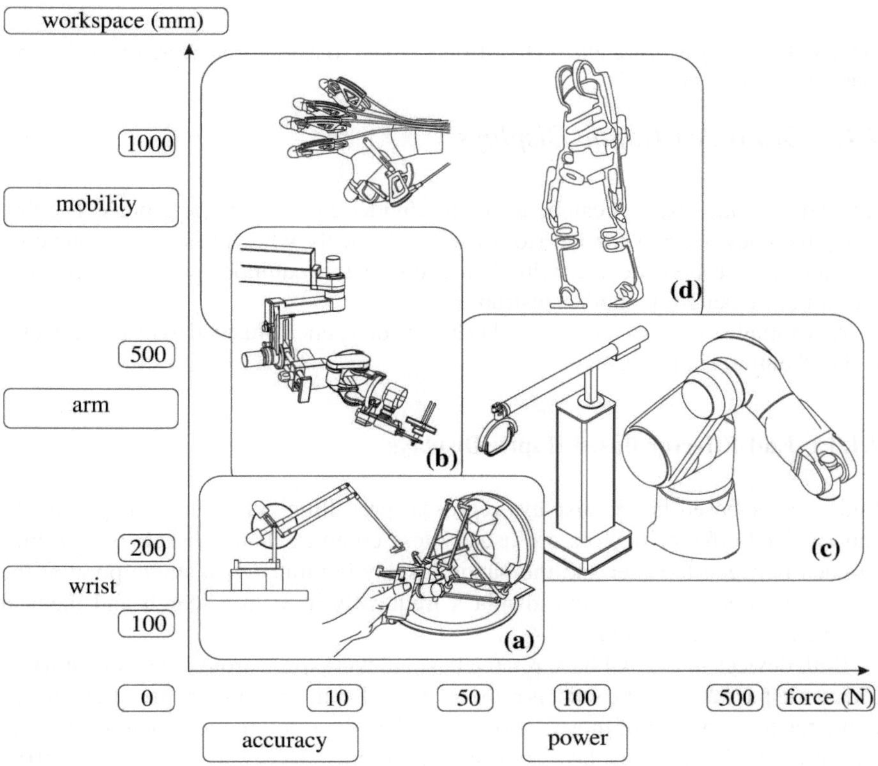

Fig. 4.3 Haptic displays classified based on their workspace, power and accuracy: **a** haptic displays for hand and wrist, **b** arm exoskeletons, **c** haptic displays based on industrial manipulators, **d** mobile haptic displays

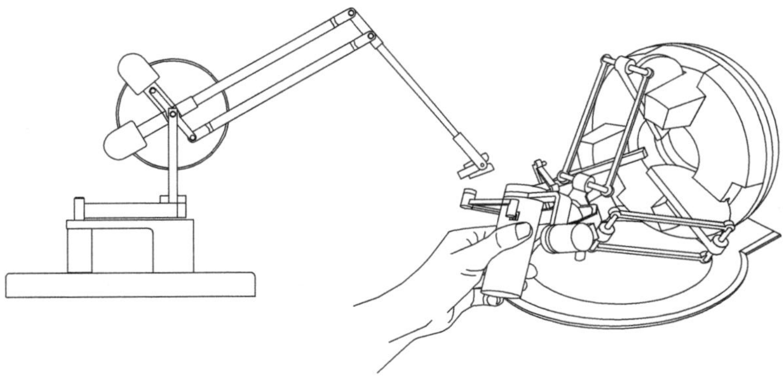

Fig. 4.4 Two examples of end-effector based haptic displays (Phantom from Sensable and Omega.7 from Force Dimension)

4.1.3 Grounded Haptic Displays

Grounded haptic displays can be generally divided into two major groups. The first group includes steering handles (joysticks) and wheels, while the second group consists mostly of robotic devices with their end-effectors equipped with tools, either in the form of a pencil or another instrument.

An example of a device with a tool in a form of a pencil attached to its end-effector is the Phantom haptic display.

4.1.3.1 End-Effector Based Haptic Displays

End-effector based haptic displays represent prevailing concepts for haptic applications. End-effector devices are usually less complex than exoskeleton systems. Contact between the user and the haptic display is limited to a single point at the robot end-effector. Two haptic displays frequently used in research and medical applications are shown in Fig. 4.4.

Both devices in Fig. 4.4 have relatively small workspace around the human wrist. The workspace around the wrist is relevant, since haptic interactions are often limited to the range of motion of fingers and wrist, when the forearm movement is partially constrained. The two displays also produce limited forces in the range of 10 N. Both devices enable very precise manipulations. However, they are different in their kinematic structures (Phantom is based on a series mechanism concept, while Omega is a parallel mechanism), as well as dynamic properties.

Both displays are distinguished by low mass (the Phantom device weight is mostly passively compensated, while the Omega device requires active gravity compensation), low friction, high rigidity and backdrivability. Therefore, the two displays enable generation of convincing impressions of contact, restricted movement, surface compliance, surface friction, textures and other mechanical properties of virtual objects.

The user is coupled to the mechanism either by inserting the finger into a thimble or through the model of a stylus (Phantom), or by grasping the haptic display end-effector (Omega). The two devices were designed based on certain assumptions that can be summarized as follows [8].

An important component of human capabilities of visualization, memorization and building of cognitive models of the physical environment, originates from haptic interactions with objects in the environment. Kinesthetic and tactile senses and perception of force, together with motor capabilities, enable exploration, detection and manipulation of objects in a physical environment. Information about object movement in relation to the applied force and forces required for displacing the object allow estimation of geometric cues (form, position), mechanical characteristics (impedance, friction, texture) and events (constraints, variations, touch, slip).

Most haptic interactions with the environment involve small or zero torques at the fingertip while in contact with the environment. Therefore, a haptic display with three active degrees of freedom enables simulation of a large number of different haptic tasks. A device should also allow unconstrained movement in a free space. Therefore, it should not apply forces on the user when movement is performed in free space. This means that the device should have small intrinsic friction, small inertia and no unbalanced mass.

One of the criteria that determines the quality of a haptic interface is the maximal stiffness that the interface is capable of displaying to the user. Since neither the mechanism nor the control algorithm are infinitely stiff, the maximal stiffness of virtual objects that can be rendered, depends mainly on the stiffness of the control system. In a virtual environment the wall surfaces should feel stiff. This means that device actuators should not saturate too quickly when in contact with virtual constraints.

Due to its simple kinematic structure, Phantom haptic display kinematics will be analyzed in more details in the following paragraphs. The display actually represents a link between the direct current motors equipped with position encoders and the user's finger. The spatial position of the finger can be measured using the position encoders, while device actuators generate spatial forces on the finger. Motor torques are transmitted across pretensioned tendons to a lightweight mechanism. At the mechanism end-effector a passive or active thimble with three degrees of freedom is attached. Passive thimble axes intersect at a single point and in that point there is no torque, only force. This allows placing of a finger into an arbitrary orientation. An active thimble enables generation of torques at the display end-effector.

Figure 4.5 shows the Phantom haptic display in a reference pose. The base coordinate frame is not located at the robot's base, but rather at a location where it is aligned with the end-effector coordinate frame, when the device is in its reference pose. We will assume a haptic display with three active degrees of freedom (no torque at end-effector), which are characterized by three joint variables combined in a vector $\mathbf{q} = \begin{bmatrix} q_1 & q_2 & q_3 \end{bmatrix}^T$.

The device forward kinematics will be computed using vector parameters [9]. If we assume placement of coordinate frames as shown in Fig. 4.5, we can determine vector

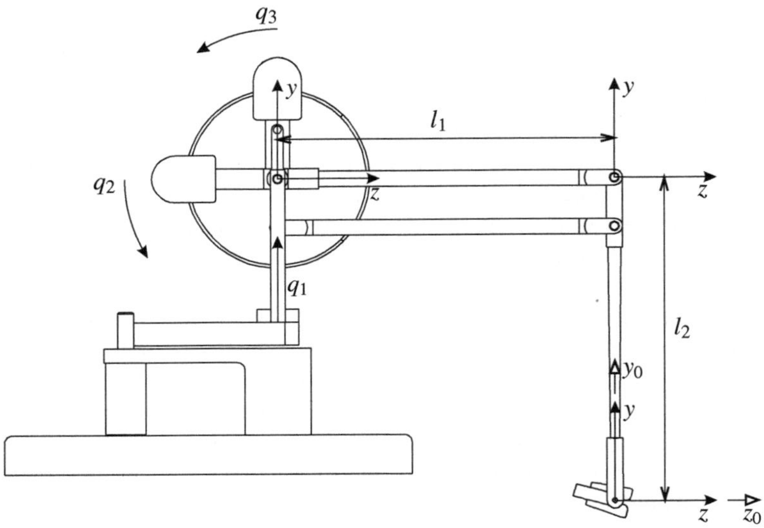

Fig. 4.5 Reference pose of the Phantom haptic display

Table 4.1 Vector parameters and joint variables for the Phantom haptic display

i	1	2	3
q_i	q_1	q_2	$(q_3 - q_2)$

i	1	2	3
$\mathbf{e}_i$	0	-1	-1
	1	0	0
	0	0	0

i	1	2	3	4
$\mathbf{b}_{i-1}$	0	0	0	0
	0	0	0	$-l_2$
	0	0	l_1	0

parameters and joint variables for the device mechanism as defined in Table 4.1. Due to the parallel mechanism used for actuating the third joint, the joint angle is defined as the difference $(q_3 - q_2)$ (see Fig. 4.6 for details).

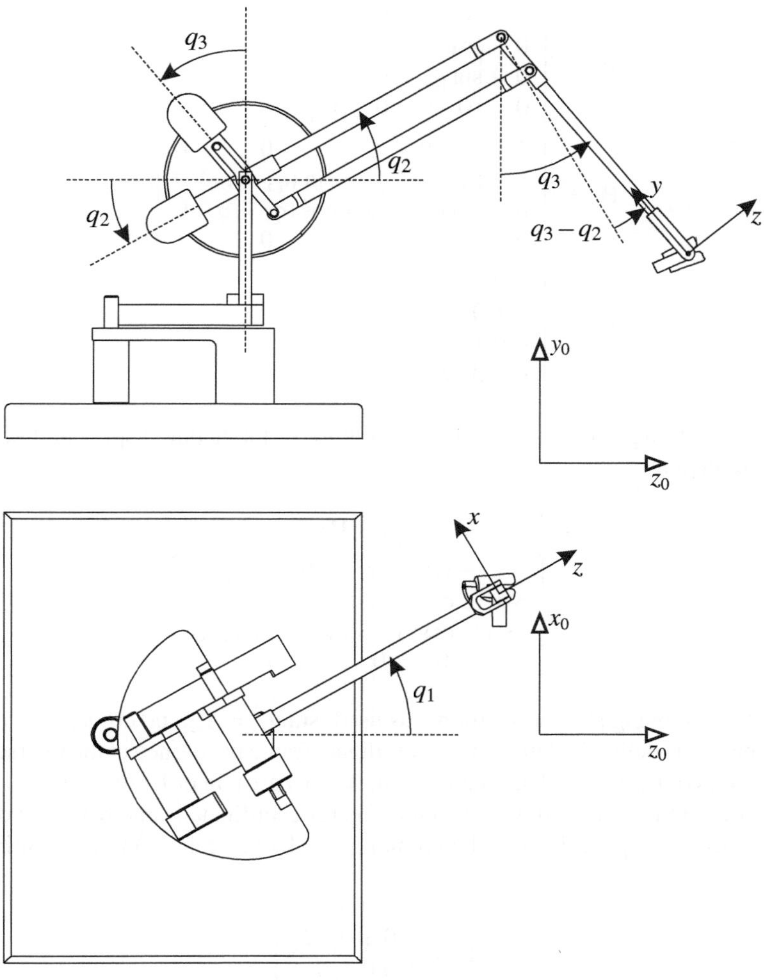

Fig. 4.6 *Front* and *top* view of the Phantom haptic display

The selected vector parameters of the mechanism are written into the homogenous transformation matrices

$$
{}^{0}\mathbf{H}_1 = \begin{bmatrix} \cos q_1 & 0 & -\sin q_1 & 0 \\ 0 & 1 & 0 & 0 \\ -\sin q_1 & 0 & \cos q_1 & 0 \\ 0 & 0 & 0 & 1 \end{bmatrix}, \tag{4.2}
$$

$$ {}^{1}\mathbf{H}_2 = \begin{bmatrix} 1 & 0 & 0 & 0 \\ 0 & \cos q_2 & \sin q_2 & 0 \\ 0 & -\sin q_2 & \cos q_1 & 0 \\ 0 & 0 & 0 & 1 \end{bmatrix}, \tag{4.3} $$

$$ {}^{2}\mathbf{H}_3 = \begin{bmatrix} 1 & 0 & 0 & 0 \\ 0 & \cos(q_3 - q_2) & \sin(q_3 - q_2) & 0 \\ 0 & -\sin(q_3 - q_2) & \cos(q_3 - q_2) & 0 \\ 0 & 0 & 0 & 1 \end{bmatrix}, \tag{4.4} $$

$$ {}^{3}\mathbf{H}_4 = \begin{bmatrix} 1 & 0 & 0 & 0 \\ 0 & 1 & 0 & -l_2 \\ 0 & 0 & 1 & 0 \\ 0 & 0 & 0 & 1 \end{bmatrix}. \tag{4.5} $$

By multiplying matrices (4.2)–(4.5), the pose of the haptic display end-effector can be written as

$$ \begin{aligned} \mathbf{T}_b &= {}^{0}\mathbf{H}_1 \cdot {}^{1}\mathbf{H}_2 \cdot {}^{2}\mathbf{H}_3 \cdot {}^{3}\mathbf{H}_4 \\ &= \begin{bmatrix} c_1 & -s_1 s_3 & c_3 s_1 & s_1(l_1 c_2 + l_2 s_3) \\ 0 & c_3 & s_3 & -l_2 c_3 + l_1 s_2 \\ -s_1 & -c_1 s_3 & c_1 c_3 & c_1(l_1 c_2 + l_2 s_3) \\ 0 & 0 & 0 & 1 \end{bmatrix}, \end{aligned} \tag{4.6} $$

where the following shorter notation was used: $\sin\vartheta_\angle = s_\angle$ and $\cos\vartheta_\angle = c_\angle$. The above matrix is defined relative to the coordinate frame positioned at the intersection of joint axes of the first and the second joint (robot base). In order to transform it to a location where it is aligned with the end-effector coordinate frame when the device is in the reference pose, it has to be premultiplied by the transformation matrix

$$ {}^{b}\mathbf{H}_e = \begin{bmatrix} 1 & 0 & 0 & 0 \\ 0 & 1 & 0 & -l_2 \\ 0 & 0 & 1 & l_1 \\ 0 & 0 & 0 & 1 \end{bmatrix} \tag{4.7} $$

leading to

$$ \mathbf{T} = \begin{bmatrix} c_1 & -s_1 s_3 & c_3 s_1 & s_1(l_1 c_2 + l_2 s_3) \\ 0 & c_3 & s_3 & l_2 - l_2 c_3 + l_1 s_2 \\ -s_1 & -c_1 s_3 & c_1 c_3 & -l_1 + c_1(l_1 c_2 + l_2 s_3) \\ 0 & 0 & 0 & 1 \end{bmatrix}. \tag{4.8} $$

End-effector orientation matrix $\mathbf{R}$ is a submatrix of (4.8)

$$ \mathbf{R} = \begin{bmatrix} c_1 & -s_1 s_3 & c_3 s_1 \\ 0 & c_3 & s_3 \\ -s_1 & -c_1 s_3 & c_1 c_3 \end{bmatrix}. \tag{4.9} $$

Often it is also necessary to solve the inverse kinematics of the display mechanism. The inverse kinematics problem means computation of joint angles (q_1, q_2, q_3) as a function of end-effector position $\mathbf{p} = \begin{bmatrix} p_x & p_y & p_z \end{bmatrix}^T$ [10]. Based on relations in Fig. 4.6 we can compute the first joint angle q_1 as

$$q_1 = \arctan 2(p_x, p_z + l_1), \tag{4.10}$$

where arctan 2 is a four-quadrant arctangent function. Angles q_2 in q_3 can be computed from relations in Fig. 4.7. First we calculate distances R and r

$$R = \sqrt{p_x^2 + (p_z + l_1)^2}, \tag{4.11}$$

$$r = \sqrt{p_x^2 + (p_y - l_2)^2 + (p_z + l_1)^2}. \tag{4.12}$$

Angle $\beta = \angle(R, r)$ then equals

$$\beta = \arctan 2(p_y - l_2, R). \tag{4.13}$$

By using the cosine law for the triangle $\Delta(l_1, l_2, r)$ we can compute angle γ as

$$l_1^2 + r^2 - 2l_1 r \cos\gamma = l_2^2, \tag{4.14}$$

$$\gamma = \arccos \frac{l_1^2 + r^2 - l_2^2}{2l_1 r}. \tag{4.15}$$

The workspace of the device is limited such that $\gamma > 0$, therefore

$$q_2 = \gamma + \beta. \tag{4.16}$$

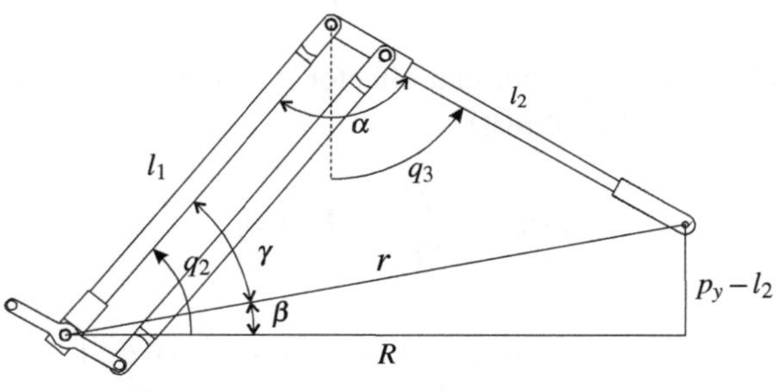

Fig. 4.7 Parallelogram mechanism of the Phantom haptic display

In order to compute angle q_3, we again write the cosine law for the triangle $\triangle(l_1, l_2, r)$, only this time by taking into account angle α

$$l_1^2 + l_2^2 - 2l_1 l_2 \cos\alpha = r^2, \tag{4.17}$$

$$\alpha = \arccos \frac{l_1^2 + l_2^2 - r^2}{2l_1 l_2}. \tag{4.18}$$

Angle α is always positive in the workspace of the device, thus we can write

$$q_3 = q_2 + \alpha - \frac{\pi}{2}. \tag{4.19}$$

Finally, the next few paragraphs will focus on analysis of differential kinematics of the haptic device and computation of the Jacobian matrix that can be used to transform velocities, forces and torques between the task space coordinates and joint variables. First end-effector orientation is expressed in terms of RPY angles $\boldsymbol{\phi} = \begin{bmatrix} \varphi \ \vartheta \ \psi \end{bmatrix}^T$ from rotation matrix $\mathbf{R}(\mathbf{q})$,

$$\mathbf{R}(\mathbf{q}) = \mathbf{R}(\boldsymbol{\phi}) = \mathbf{R}_z(\boldsymbol{\varphi})\mathbf{R}_y(\boldsymbol{\vartheta})\mathbf{R}_x(\boldsymbol{\psi}) = \begin{bmatrix} c_\varphi c_\vartheta & c_\varphi s_\vartheta s_\psi - s_\varphi c_\psi & c_\varphi s_\vartheta c_\psi + s_\varphi s_\psi \\ s_\varphi c_\vartheta & s_\varphi s_\vartheta s_\psi + c_\varphi c_\psi & s_\varphi s_\vartheta c_\psi - c_\varphi s_\psi \\ -s_\vartheta & c_\vartheta s_\psi & c_\vartheta c_\psi \end{bmatrix}, \tag{4.20}$$

where $c_\angle$ represents $\cos(\angle)$ and $s_\angle$ represents $\sin(\angle)$. By comparing Eqs. (4.9) and (4.20) *RPY* angles can be determined as

$$\begin{aligned} c_1 = c_\varphi c_\vartheta \quad \wedge \quad 0 = s_\varphi c_\vartheta \quad &\Rightarrow \quad \varphi = 0, \\ -s_1 = -s_\vartheta \quad &\Rightarrow \quad \vartheta = q_1, \\ -c_1 s_3 = c_\vartheta s_\psi \quad \wedge \quad c_1 c_3 = c_\vartheta c_\psi \quad &\Rightarrow \quad \psi = -q_3 \end{aligned} \tag{4.21}$$

or

$$\boldsymbol{\phi} = \begin{bmatrix} 0 \ q_1 \ -q_3 \end{bmatrix}^T. \tag{4.22}$$

Translational velocity of the device end-effector is computed as a time derivative of the end-effector position vector $\mathbf{p}(\mathbf{q})$

$$\dot{\mathbf{p}}(\mathbf{q}) = \frac{\partial \mathbf{p}}{\partial \mathbf{q}}\dot{\mathbf{q}} = \mathbf{J}_P(\mathbf{q})\dot{\mathbf{q}}, \tag{4.23}$$

where

$$\mathbf{J}_P(\mathbf{q}) = \begin{bmatrix} c_1(l_1 c_2 + l_2 s_3) & -l_1 s_1 s_2 & l_2 s_1 c_3 \\ 0 & l_1 c_2 & l_2 s_3 \\ -s_1(l_1 c_2 + l_2 s_3) & -l_1 c_1 s_2 & l_2 c_1 c_3 \end{bmatrix}. \tag{4.24}$$

Rotational velocity of the end effector is computed as a time derivative of the end-effector orientation $\boldsymbol{\phi}(\mathbf{q})$,

$$\dot{\boldsymbol{\phi}}(\mathbf{q}) = \frac{\partial \boldsymbol{\phi}}{\partial \mathbf{q}} \dot{\mathbf{q}} = \mathbf{J}_{\phi}(\mathbf{q})\dot{\mathbf{q}}, \tag{4.25}$$

where

$$\mathbf{J}_{\phi}(\mathbf{q}) = \begin{bmatrix} 0 & 0 & 0 \\ 1 & 0 & 0 \\ 0 & 0 & -1 \end{bmatrix}. \tag{4.26}$$

By combining translational and rotational relations we obtain the analytical Jacobian matrix as

$$\mathbf{J}_A(\mathbf{q}) = \begin{bmatrix} \mathbf{J}_P(\mathbf{q}) \\ \mathbf{J}_{\phi}(\mathbf{q}) \end{bmatrix} = \begin{bmatrix} c_1(l_1c_2 + l_2s_3) & -l_1s_1s_2 & l_2s_1c_3 \\ 0 & l_1c_2 & l_2s_3 \\ -s_1(l_1c_2 + l_2s_3) & -l_1c_1s_2 & l_2c_1c_3 \\ 0 & 0 & 0 \\ 1 & 0 & 0 \\ 0 & 0 & -1 \end{bmatrix}. \tag{4.27}$$

4.1.3.2 Exoskeleton Based Haptic Displays

An arm exoskeleton haptic display is a device that measures user's arm movements and applies desired forces onto the arm. It enables display of dynamic properties of virtual objects and forces resulting from collisions of any part of the extremity with a virtual environment.

An example of an arm exoskeleton (ARMin, [11]) with six active degrees of freedom (three for the shoulder, one for the elbow and two for the wrist) is shown in Fig. 4.8. In addition to the six active degrees of freedom, the display consists of additional passive degrees of freedom for comfortable adjustment to different arm anthropometric properties.

The haptic display enables display of forces and torques acting on the upper extremity while interacting with objects within a virtual environment. Due to its exoskeleton structure, it allows precise application of forces and torques on individual arm joints and not only on the hand as it is the case with end-effector based haptic devices. The exoskeleton haptic display uniquely defines poses of all arm segments.

The ARMin haptic display was primarily developed for assistance in neuromotor rehabilitation. A person with motor disorders can practice arm movements in a virtual environment. However, the haptic display does not only simulate contacts with the virtual environment, but it can also actively assist in moving the disabled extremity.

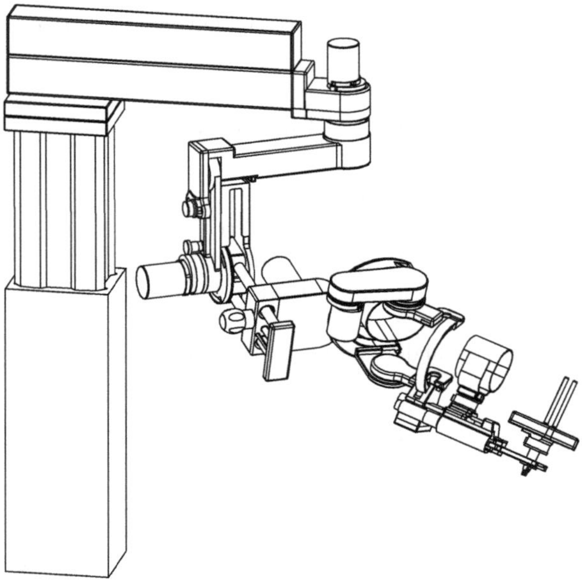

Fig. 4.8 Haptic exoskeleton ARMin (ETH Zurich) with six active degrees of freedom

4.1.4 Mobile Haptic Displays

Mobile haptic displays are devices that fit on user's limbs and move along with the user. Since they are kinematically similar to legs, arms, hands and fingers, of which the movement they measure and to which they apply forces, they achieve the largest possible workspace.

Mobile haptic devices are mostly designed for lower extremities, where they are predominantly used as *human power amplifiers*, or hands. Though similar concepts apply to human power amplifiers as for haptic devices, these are not of primary interest of this book. Therefore, a hand exoskeleton device will be addressed as an example. The biggest obstacle in the design of haptic displays for the hand is caused by a high complexity of human hands, where 22 degrees of freedom are constrained in a small space.

A haptic display for the hand applies forces on fingers while the device is usually attached to the forearm. The device has the shape of a glove, where actuators are placed on the forearm and actuation forces are transmitted to the fingers across tendons and pulleys. An example of such a device is CyberGrasp haptic display (Fig. 4.9) that provides force feedback to individual fingers. The device is active only during finger flexion by pulling on the tendons attached to individual fingers across an exoskeleton mechanism. The device's five actuators, one for each finger, are placed on the forearm.

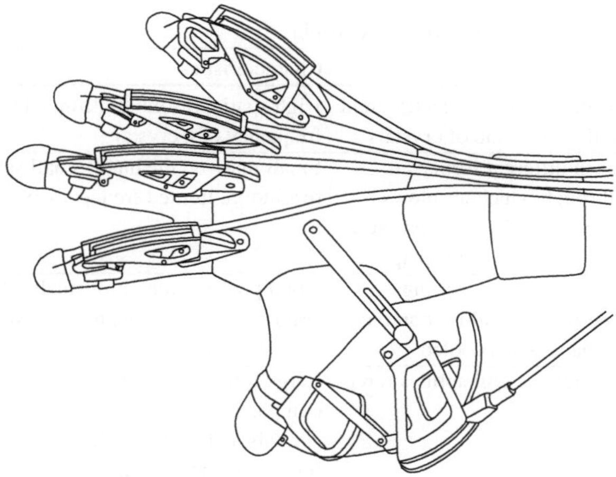

Fig. 4.9 CyberGrasp (CyberGlove Systems) haptic display

The CyberGrasp displays forces during grasping. These forces are approximately perpendicular to the fingertips in the entire workspace of fingers. Grasping force can be set individually for each finger. Implementation of attachment of tendons on the thumb allows for full closure of the fist.

4.2 Tactile Haptic Displays

Kinesthetic haptic displays are suitable for relatively coarse interactions with virtual objects. For the purpose of precise rendering of contact parameters within the contact region, tactile displays must be used. Tactile sensing plays an important role during manipulation and discrimination of objects, where force sensing is not efficient enough. Sensations are important for assessment of local shape, texture and temperature of objects as well as for detecting slippage. Tactile senses also provide information about compliance, elasticity and viscosity of objects. Sensing of vibrations is important for detection of the objects textures as well as for measuring vibrations. At the same time it also shortens reaction times and minimizes contact forces. Since reaction force is not generated prior to object deformation, tactile information becomes relevant also for the initial contact detection. This significantly increases abilities for detecting contacts, measuring contact forces and tracking a constant contact force. Finally, tactile information is also necessary for minimizing interaction forces in tasks that require precise manipulations.

In certain circumstances a tactile display of one type can be replaced with a display of another type. A temperature display can, for example, be used for simulating object material properties.

Table 4.2 Actuation technologies for tactile display

Actuation	Characteristics
Piezoelectric crystal: changes in electric field cause expansion and contraction of crystals	High spatial resolution; limited to the resonant frequency of a crystal
Pneumatics comes in different forms: an array of air holes through which air blows, matrix of small bubbles in which the pressure changes, a bubble in a form of a fingertip	Small mass on the hand; small spatial and temporal resolution, limited frequency bandwidth
Shape-memory alloy: wires and springs from different shape-memory alloys contract when heated and expand when cooled	Good power/force ratio; small efficiency when contracting, slow heat transmission limits wire relaxation time
Electromagnet: a magnetic coil applies force on a metal piston	Large force in a steady state; better bandwidth compared to other materials (except piezoelectric crystals and sound coils); relatively high mass, nonlinear, therefore control more challenging
Sound coil: coil transmits vibrations of different frequencies onto the skin	High temporal resolution, relatively small, thus it does not affect natural fingers movement; limited spatial resolution, limited scaling
Heat pump: device transfers energy either by heating or cooling the skin	Does not require any fluids; limited temporal and spatial resolution, size, limited bandwidth

Tactile stimulation can be achieved using different approaches. Systems that are most often used in virtual environments include mechanical needles actuated using electromagnets, piezoelectric crystals or shape-memory alloy materials, vibrators, that are based on sound coils, pressure from pneumatic systems or heat pumps. Main characteristics of these approaches are summarized in Table 4.2 [12].

References

1. Youngblut, C., Johnson, R.E., Nash, S.H., Wienclaw, R.A., Will, C.A.: Review of virtual environment interface technology. Ida paper p-3786, Institute for Defense Analysis, Virginia, USA (1996)
2. Burdea, G.: Force and Touch Feedback for Virtual Reality. Wiley, New York (1996)
3. Hollerbach, J.M.: Some current issues in haptics research. In: Proceedings of the IEEE International Conference on Robotics and Automation, pp. 757–762 (2000)
4. Hannaford, B., Venema, S.: Kinesthetic displays for remote and virtual environments. Virtual Environments and Advanced Interface Design, pp. 415–436. Oxford University Press, New York (1995)
5. Bar-Cohen, Y.: Automation, Miniature Robotics and Sensors for Non-Destructive Testing and, Evaluation. American Society for Nondestructive Testing (2000)
6. Hayward, V., Astley, O.R.: Performance measures for haptic interfaces. In: The 7th International Symposium on Robotics Research, pp. 195–207 (1996)

7. Richard, C., Okamura, A., Cutkosky, M.C.: Getting a feel for dynamics: using haptic interface kits for teaching dynamics and control. Proceedings of the 1997 ASME IMECE 6th Annual Symposium on Haptic Interfaces, Dallas, TX. USA, pp. 15–25 (1997)
8. Massie, T.H., Salisubry, J.K.: The Phantom haptic interface: a device for probing virtual objects. In: Haptic Interfaces for Virtual Environment and Teleoperator Systems, Chicago, pp. 295–301 (1994)
9. Lenarčič, J.: Kinematics. The International Encyclopedia of Robotics. Wiley, New York (1988)
10. Cavusoglu, M.C., Sherman, A., Tendick, F.: Design of bilateral teleoperation controllers for haptic exploration and telemanipulation of soft environments. IEEE Trans. Robot. Autom. **18**, 641–647 (2002)
11. Nef, T., Mihelj, M., Riener, R.: Armin: a robot for patient-cooperative arm therapy. Med. Biol. Eng. Comput. **45**, 887–900 (2007)
12. Hasser, C.: Tactile feedback for a force-reflecting haptic display. Technical report, Armstrong Lab Wright-Patterson Afb Oh Crew Systems Directorate (1995)

Chapter 5
Collision Detection

The algorithm for haptic interaction with a virtual environment consists of a sequence of two tasks. When the user operates a virtual tool attached to a haptic interface, the new tool pose is computed and possible collisions with objects in a virtual environment are determined. In case of a contact, reaction forces are computed based on the environment model and force feedback is provided to the user via the haptic display. Collision detection guarantees that objects do not float into each other. A special case of contact represents grasping of virtual objects as shown in Fig. 5.1 that allows object manipulation. If grasping is not adequately modeled, it might happen that the virtual hand passes through the virtual object and the reaction forces that the user perceives are not consistent with the visual information.

In haptic interactions we have to consider two types of contact. In the first case, we are dealing with collision detection in a remote (teleoperated) environment, where the haptic interface is used for controlling a remote robot manipulator (slave system) and the contact information between the slave system and the environment is presented to the operator via the haptic interface. In the second case, we are dealing with collision detection between a virtual tool, objects and surroundings in a virtual environment.

5.1 Collision Detection for Teleoperation

In a teleoperation system it is necessary to measure interaction force between the slave system and the environment and to transmit the measured force to the haptic interface. For this purpose it is possible to apply force and torque sensors or tactile sensors developed for robotic applications. Force and torque sensors can be used for measuring forces between objects as well as for measuring forces and torques acting during the robot based manipulation of objects. Tactile sensors, contrary to the force sensors, measure contact parameters within the contact area between the slave system and the object, similarly as the human tactile sensors measure contact parameters within a limited area of contact.

M. Mihelj and J. Podobnik, *Haptics for Virtual Reality and Teleoperation*,
Intelligent Systems, Control and Automation: Science and Engineering 64,
DOI: 10.1007/978-94-007-5718-9_5,

Fig. 5.1 Grasping of virtual objects

5.1.1 Force and Torque Sensors

Interaction forces acting on a robot wrist as a result of object manipulation can be measured using force and torque sensors mounted at the robot wrist. Most efficient measurements can be achieved by mounting the force sensor between the last robot segment and the tool. Such force transducers are often referred to as wrist force sensors.

5.1.2 Tactile Sensors

Tactile sensing is defined as continuous sensing of variable contact force within an area with a high temporal-spatial resolution. Tactile sensing is in general more complex than a contact perception, which is often limited to a vector measurement of force/torque in a single point. Tactile sensors attached to fingers of a robot gripper are sensitive to information such as pressure, force and distribution of force within a contact region between a gripper and an object. They can be used, for example, to determine whether an object is merely positioned at a given place or is coupled to another object. In complex assembly tasks tactile sensing provides information about geometric relationships between different objects and enables precise control of hand movements. Tactile sensors do not only collect information required for hand control, but also enable identification of size, shape and stiffness properties of objects. All this becomes relevant for computation of grasping forces when dealing with fragile objects.

Technologies for tactile sensing are based on conducting elastomers, strain gauges, piezoelectric crystals, capacitive or optoelectronic sensors. These technologies can be divided into two major groups:

- force sensitive sensors (conducting elastomers, strain gauges, piezoelectric crystals) measure primarily contact forces,
- displacement sensitive sensors (capacitive or optoelectronic sensors) measure primarily mechanical deformation of the object.

5.2 Collision Detection in a Virtual Environment

If virtual objects fly through each other, this creates a confusing visual effect, thus penetration of one object into the other needs to be prevented. When two objects try to penetrate each other, we are dealing with collision.

Detection of contact or collision between virtual objects in a computer generated virtual environment represents a completely different problem compared to detection of contact between a tool and an object in a remote (teleoperated) environment. Since there is no physical contact between objects in a virtual environment, it is necessary to build a virtual force sensor.

Collision detection is an important step toward physical modeling of a virtual environment. It includes automatic detection of interactions between objects and computation of contact coordinates. At the moment of collision the simulation generates a response to the contact. If the user is coupled to one of the virtual objects (for example, via a virtual hand), then the response to the collision results in forces, vibrations or other haptic quantities being transmitted to a user via a haptic interface.

Before dealing with methods for collision detection in virtual environments, we shall review basic concepts of geometric modeling of virtual objects, since the method for collision detection significantly depends on the object model [1–3]. Most methods for geometric modeling originate from computer graphics. Object models are often represented using the object's exterior surfaces—the problem of model representation simplifies to a mathematical model for describing the object's surface, which defines outside boundaries of an object. These representations are often referred to as *representations with boundary surface*. Other representations are based on *constructed solid geometry*, where solid objects are used as basic blocks for modeling, or *volumetric representations*, which model objects with vector fields.

Haptic rendering is in general based on completely different requirements as computer graphics—sampling frequency of a haptic system is significantly higher and haptic rendering is of a more local nature, since we cannot physically interact with the entire virtual environment at once. Haptic rendering thus constructs a specific set of techniques making use of representational models developed primarily for computer graphics.

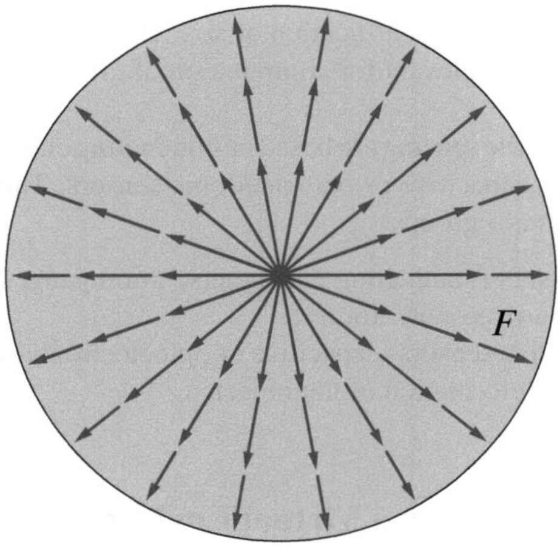

Fig. 5.2 Spherical object modeling using a force vector field method

5.2.1 Representational Models for Virtual Objects

The following section provides an overview of some modeling techniques for virtual objects with an emphasis on attributes specific for haptic collision detection.

Two early approaches for representation of virtual objects were based on a force vector field method and an intermediate plane method. The vector field corresponds to the desired reaction forces. The interior of an object is divided into areas, of which the main characteristic is the common direction of force vectors, whereas the force vector length is proportional to the distance from the surface (Fig. 5.2).

An intermediate plane [4], on the other hand, simplifies representation of objects modeled with boundary surfaces. The intermediate plane represents an approximation of the below lying object geometry with a simple planar surface. The plane parameters are refreshed as the virtual tool moves across the virtual object. However, the refresh rate of the intermediate plane can be lower than the frequency of the haptic system.

Other representational models originate from the field of computer graphics.

- *Implicit surface*
  An implicit surface is defined by the implicit function. It is defined by mapping the three-dimensional space to the space of real numbers $f : \Re^3 \mapsto \Re$ and an implicit surface is defined with points, where $f(x, y, z) = 0$. Such a function uniquely defines what is inside $f(x, y, z) < 0$ and what is outside $f(x, y, z) > 0$ of the model. Implicit surfaces are consequently generically closed surfaces.
  If function $f(x, y, z)$ is a polynomial of variables x, y and z, we are dealing with algebraic functions. A special form of algebraic function is a second-order polynomial representing cones, spheres and cylinders in a general form (Fig. 5.3).

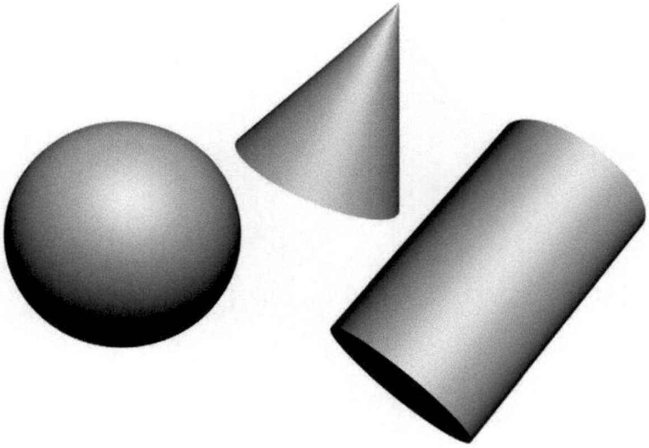

Fig. 5.3 Examples of implicit surfaces defined by second-order polynomials

- *Parametric surface*
  A parametric surface (Fig. 5.4) is defined by mapping from a subset of the plane into a three-dimensional space $f : \Re^2 \mapsto \Re^3$. Contrary to implicit surfaces, parametric surfaces are not generically closed surfaces, thus, they do not present the entire object model, but only a part of the object boundary surface.
- *Polygonal model*
  Polygonal models are most often used in computer graphics. Representations using polygons are simple, polygons are versatile and appropriate for fast geometric computations. Polygon models enable presentation of objects with boundary surfaces. An example of a polygonal model is shown in Fig. 5.5, where the most simple polygons—triangles—are used. Each object surface is represented with a triangle that is defined with three points (for example, $tr_1 = \triangle P_0 P_1 P_2$).
  Haptic rendering based on polygonal models may cause force discontinuities at the edges of individual polygons, when the force vector normal moves from the current to the next polygon. The human sensing system is accurate enough to perceive such discontinuities, meaning that these must be compensated. A method for removing discontinuities is referred to as *force shading* and is based on interpolation of normal vectors between two adjacent polygons.
  Use of polygonal models for haptic rendering is widespread, therefore collision detection based on polygonal models will be presented in more detail.

5.2.2 Collision Detection for Polygonal Models

Haptic rendering represents computation of forces required for generation of impression of a contact with the virtual object. These forces typically depend on the penetration depth of the virtual tool into the virtual object and on the direction of the tool

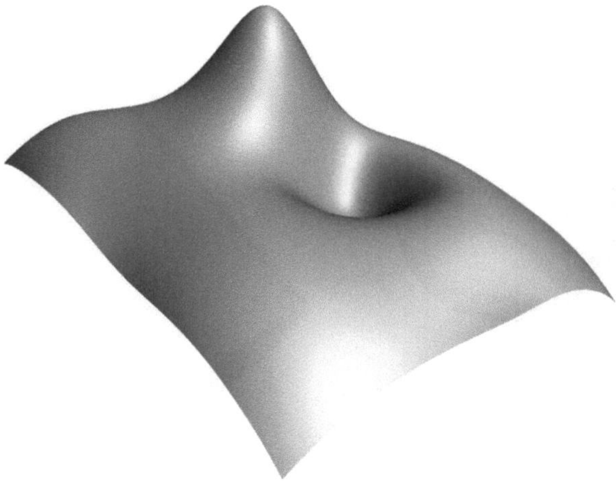

Fig. 5.4 An example of a parametric surface

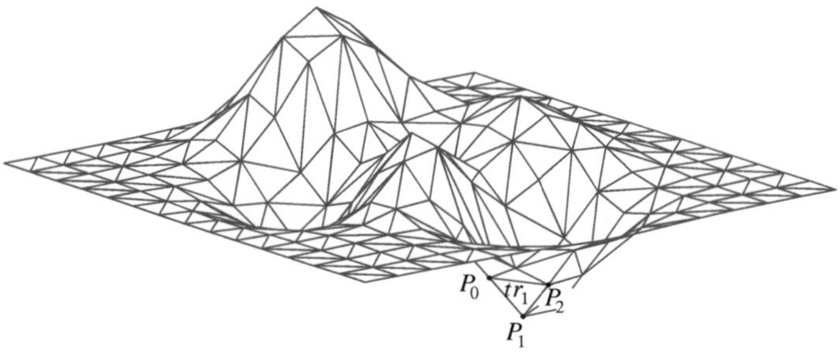

Fig. 5.5 Object modeling using triangles

acting on the object. Due to the complexity of computation of reaction forces in the case of complex environments, a virtual tool is often simplified to a point or a set of points representing the tool's endpoint. The computation of penetration depth thus simplifies to finding the point on the object that is the closest to the tool endpoint. As the tool moves, also the closest point on the object changes and must be refreshed with a sampling frequency of the haptic system.

We will analyze a simple single-point haptic interaction of a user with a virtual environment. In this case the pose of the haptic interface end-effector is measured. This is referred to as a *haptic interaction point* **HIP**. Then it is necessary to determine if the point lies within the boundaries of the virtual object. If this is the case, then the penetration depth is computed as the difference between **HIP** and the corresponding point on the object's surface (*surface contact point* **SCP**). Finally, the resulting reaction force is estimated based on the physical model of the virtual object [5]. Figure 5.6 illustrates relations during a haptic contact in a virtual environment. Points $\mathbf{P}_1$ and $\mathbf{P}_2$ define a vector $\overrightarrow{\mathbf{P}_1\mathbf{P}_2}$, that determines the intersecting ray with

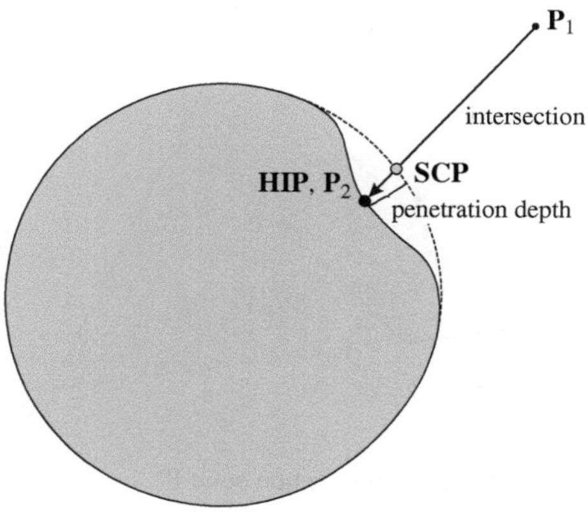

Fig. 5.6 Conditions during a penetration into the virtual object

a virtual object, in other words a ray that penetrates the object's surface. Point $\mathbf{P}_1$ is usually defined by the pose of the tool at the moment just before touching the object and point $\mathbf{P}_2$ is defined by the pose of the tool after touching the object in the next computational step. Point **HIP** equals point $\mathbf{P}_2$.

Since computation of **SCP** depends mainly on the **HIP** pose, it is clear that for a single **HIP** pose, different **SCP** points may exist, as shown in Fig. 5.7.

Next we will analyze a method of computing contact with polygonal models. The method is based on finding the intersection of a ray with a polygon [5–7]. For this purpose it is first necessary to determine whether a ray intersects the specific polygon and, if this is the case, determine the coordinates of the intersection and the parameters that define the tool position relative to the surface.

1. *Intersection of a ray and a plane containing a polygon*
   A polygon shown in Fig. 5.8 is determined by vertices $\mathbf{V}_i (i \in (0, \ldots, n-1), n \geq 3)$. Let x_i, y_i and z_i be the coordinates of the vertex $\mathbf{V}_i$. A normal vector $\mathbf{N}$

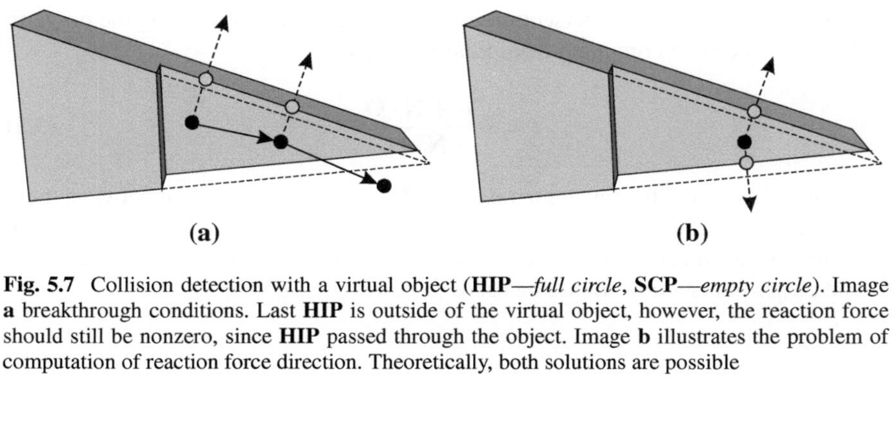

Fig. 5.7 Collision detection with a virtual object (**HIP**—*full circle*, **SCP**—*empty circle*). Image **a** breakthrough conditions. Last **HIP** is outside of the virtual object, however, the reaction force should still be nonzero, since **HIP** passed through the object. Image **b** illustrates the problem of computation of reaction force direction. Theoretically, both solutions are possible

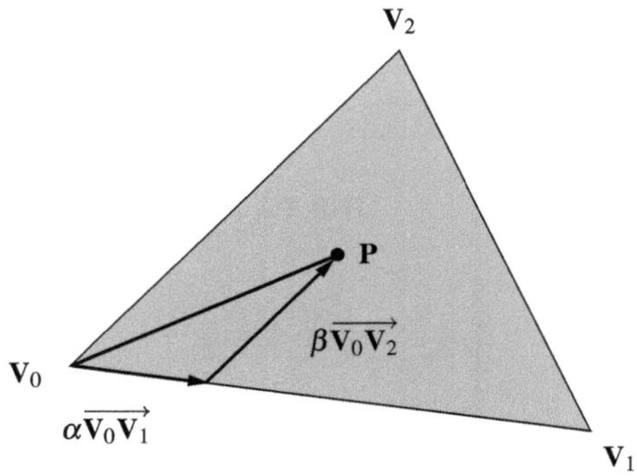

Fig. 5.8 Computation of coordinates of intersection point **P**

to the plane containing the polygon can be computed as a cross product

$$\mathbf{N} = \overrightarrow{\mathbf{V}_0\mathbf{V}_1} \times \overrightarrow{\mathbf{V}_0\mathbf{V}_2}. \tag{5.1}$$

For every point **P** lying on the plane the following relation applies:

$$(\mathbf{P} - \mathbf{V}_0) \cdot \mathbf{N} = 0. \tag{5.2}$$

Let a constant d be defined as a dot product $d = -\mathbf{V}_0 \cdot \mathbf{N}$. A general plane equation in a vector form

$$\mathbf{N} \cdot \mathbf{P} + d = 0 \tag{5.3}$$

can be computed once for each polygon and stored as a description of that polygon. Let the ray be described by a parametric vector equation as

$$\mathbf{r}(t) = \mathbf{O} + \mathbf{D}t\,, \tag{5.4}$$

where **O** defines the ray source and **D** the normalized ray direction. Parameter t that corresponds to the intersection of a ray and the plane containing the polygon ($\mathbf{r}(t) = \mathbf{P}$) can be computed from Eqs. (5.3) and (5.4) as

$$t = -\frac{d + \mathbf{N} \cdot \mathbf{O}}{\mathbf{N} \cdot \mathbf{D}}. \tag{5.5}$$

If the polygon and the ray are parallel ($\mathbf{N} \cdot \mathbf{D} = 0$), the intersection between them does not exist, if the intersection lies behind the ray source ($t \leq 0$), the intersection is not relevant and if an intersection with a closer polygon was previously detected, the closer intersection should be used.

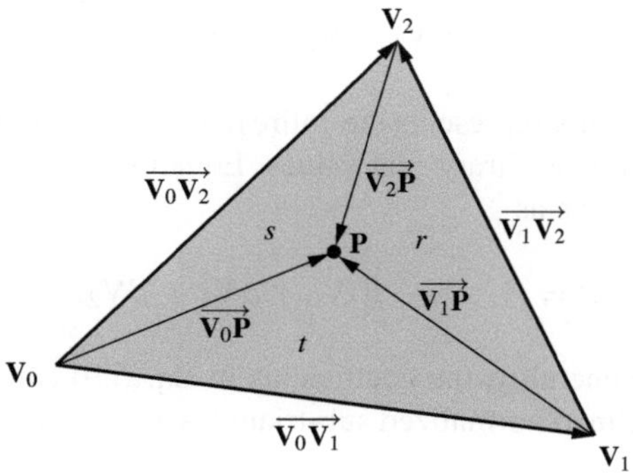

Fig. 5.9 Breakdown of a *triangle* in *sub-triangles* in relation to point **P**

2. *Computation of the location of the intersection of a ray and a plane relative to the selected polygon*
   We will analyze the computation of the intersection for the simplest polygons—triangles (Fig. 5.8) [8, 9]. If a polygon has n vertices ($n > 3$), it can be represented as a set of $n - 2$ triangles.
   Figure 5.9 shows a triangle $\triangle \mathbf{V}_0\mathbf{V}_1\mathbf{V}_2$. Point **P** lies within the boundaries of the triangle and line segments $\overline{\mathbf{V_0P}}$, $\overline{\mathbf{V_1P}}$ and $\overline{\mathbf{V_2P}}$ split the triangle into three sub-triangles, of which the sum of surface areas equals the area $A_{\triangle \mathbf{V}_0\mathbf{V}_1\mathbf{V}_2}$ of the triangle $\triangle \mathbf{V}_0\mathbf{V}_1\mathbf{V}_2$. We define areas of sub-triangles in relation to the area $A_{\triangle \mathbf{V}_0\mathbf{V}_1\mathbf{V}_2}$ as

$$\begin{aligned} A_{\triangle \mathbf{P}\mathbf{V}_1\mathbf{V}_2} &= rA_{\triangle \mathbf{V}_0\mathbf{V}_1\mathbf{V}_2}, \\ A_{\triangle \mathbf{V}_0\mathbf{P}\mathbf{V}_2} &= sA_{\triangle \mathbf{V}_0\mathbf{V}_1\mathbf{V}_2}, \\ A_{\triangle \mathbf{V}_0\mathbf{V}_1\mathbf{P}} &= tA_{\triangle \mathbf{V}_0\mathbf{V}_1\mathbf{V}_2}. \end{aligned} \tag{5.6}$$

   Since quantities r, s and t actually represent normalized sub-triangles areas, we can write the following conditions:

$$\begin{aligned} &0 \le \{r,\ s,\ t\} \le 1, \\ &r + s + t = 1. \end{aligned} \tag{5.7}$$

   The first condition in (5.7) specifies that the sub-triangles areas are positive values and each sub-triangle area is smaller than or equal to the area of the triangle $\triangle \mathbf{V}_0\mathbf{V}_1\mathbf{V}_2$. The second condition states that the sum of sub-triangles areas equals the area of triangle $\triangle \mathbf{V}_0\mathbf{V}_1\mathbf{V}_2$. If conditions (5.7) are met, point **P** lies within the triangle, otherwise it lies outside the triangle $\triangle \mathbf{V}_0\mathbf{V}_1\mathbf{V}_2$. It is trivial to verify that by assuming positive values of sub-triangles areas, the sum of these areas would be larger than the area $A_{\triangle \mathbf{V}_0\mathbf{V}_1\mathbf{V}_2}$, if the point **P** would lie outside of the boundaries of triangle $\triangle \mathbf{V}_0\mathbf{V}_1\mathbf{V}_2$.
   Next we write the parametric equation for point **P**, lying on a plane defined by points $\mathbf{V_0}$, $\mathbf{V_1}$ and $\mathbf{V_2}$ (Fig. 5.8) as

$$\mathbf{P} = \mathbf{V}_0 + \alpha(\mathbf{V}_1 - \mathbf{V}_0) + \beta(\mathbf{V}_2 - \mathbf{V}_0). \tag{5.8}$$

Equation (5.8) actually represents the entire plane, if we presume that parameters α and β can assume arbitrary real values. Equation (5.8) can be rewritten in a slightly different form as

$$\mathbf{P} = (1 - \alpha - \beta)\mathbf{V}_0 + \alpha\mathbf{V}_1 + \beta\mathbf{V}_2. \tag{5.9}$$

Without loss of generality, the coefficients in Eq. (5.9) can be substituted with the previously defined normalized sub-triangles areas r, s and t

$$\alpha = s \quad \beta = t \quad 1 - \alpha - \beta = r, \tag{5.10}$$

thus

$$\mathbf{P} = r\mathbf{V}_0 + s\mathbf{V}_1 + t\mathbf{V}_2. \tag{5.11}$$

We already determined that point $\mathbf{P}$ lies within the triangle $\triangle\mathbf{V}_0\mathbf{V}_1\mathbf{V}_2$, if conditions in (5.7) are satisfied. Based on definition (5.10), the second condition is trivially satisfied. The first condition is met if

$$\alpha \geq 0, \ \beta \geq 0 \ \text{ and } \ \alpha + \beta \leq 1. \tag{5.12}$$

We must now compute parametric coordinates α and β to be able to verify whether the intersection of a ray and a plane lies within the triangle. Equation (5.8) has three components

$$\begin{cases} x_P - x_0 = \alpha(x_1 - x_0) + \beta(x_2 - x_0) \\ y_P - y_0 = \alpha(y_1 - y_0) + \beta(y_2 - y_0) \\ z_P - z_0 = \alpha(z_1 - z_0) + \beta(z_2 - z_0). \end{cases} \tag{5.13}$$

Since intersection $\mathbf{P}$ lies on the plane determined by points $\mathbf{V}_0$, $\mathbf{V}_1$ and $\mathbf{V}_2$, there exist a unique solution for parametric coordinates (α, β). A simplification of a system of Eq. (5.13) can be achieved by projecting the triangle $\triangle\mathbf{V}_0\mathbf{V}_1\mathbf{V}_2$ to one of the basic planes, either $x - y$, $x - z$ or $y - z$. If the polygon is perpendicular to one of these basic planes, the polygon projection to that plane will result in a line segment. In order to avoid such degeneration and to guarantee the largest possible projection, we first compute the dominant axis of the normal vector to the polygon. Then we project the polygon to the plane perpendicular to that axis. For example, if z is the dominant axis of the normal vector, the polygon should be projected onto the $x - y$ plane.

Let (u, v) be the coordinates of a two-dimensional vector in this plane. Coordinates of vectors $\overrightarrow{\mathbf{V}_0\mathbf{P}}$, $\overrightarrow{\mathbf{V}_0\mathbf{V}_1}$ and $\overrightarrow{\mathbf{V}_0\mathbf{V}_2}$ projected to this plane are

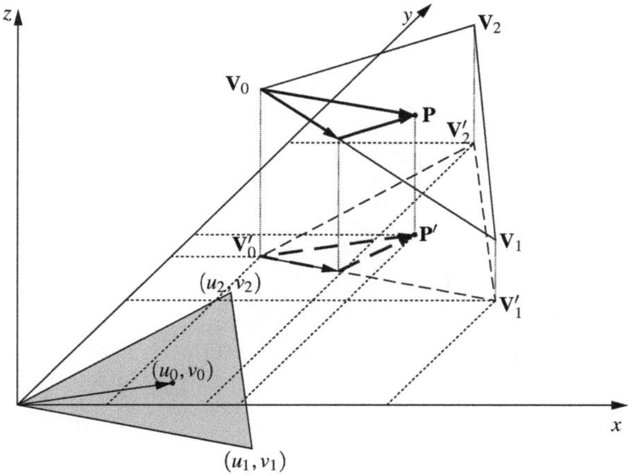

Fig. 5.10 Projection of a *polygon* on $x - y$ plane

$$\begin{array}{lll} u_0 = x_P - x_0 & u_1 = x_1 - x_0 & u_2 = x_2 - x_0 \\ v_0 = y_P - y_0 & v_1 = y_1 - y_0 & v_2 = y_2 - y_0. \end{array} \tag{5.14}$$

Relations for projection on $x - y$ plane are illustrated in Fig. 5.10. Equation (5.13) simplifies to

$$\begin{cases} u_0 = \alpha u_1 + \beta u_2 \\ v_0 = \alpha v_1 + \beta v_2. \end{cases} \tag{5.15}$$

In addition to computational efficiency, the presented method has other advantages. For example, in the case that the point **P** lies outside the triangle, parametric coordinates (α, β) still define the tool position relative to the tested triangle. From those coordinate values it is possible to define six regions surrounding the triangle. By using the results of the intersection computation, we can identify the region containing the one we seek (the one that contains the intersection) triangle. For this purpose we define a third parametric coordinate γ

$$\gamma = r = 1 - \alpha - \beta. \tag{5.16}$$

The simplest example is when one of the parametric coordinates is negative. In this case, it makes sense to take the neighboring triangle for the next test. If two parametric coordinates are negative, it is necessary to check triangles that share a single vertex with the tested triangle. Conditions are shown in Fig. 5.11.

As previously mentioned, haptic rendering of polygonal models causes force discontinuities on the edges of individual polygons. This can be corrected using the method of force shading. Using the parametric coordinates it is possible to compute the interpolated normal in point **P**

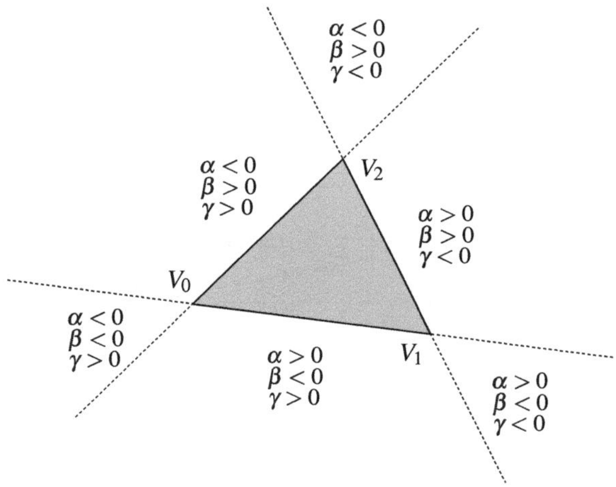

Fig. 5.11 Regions outside the *triangle* can be defined using the parametric coordinates α, β, γ

$$\mathbf{N}_P = (1 - (\alpha + \beta))\mathbf{N}_0 + \alpha\mathbf{N}_1 + \beta\mathbf{N}_2, \tag{5.17}$$

that enables smooth transitions between individual polygons. Normals $\mathbf{N}_0, \mathbf{N}_1, \mathbf{N}_2$ in vertices of triangle $\triangle\mathbf{V}_0\mathbf{V}_1\mathbf{V}_2$ are computed as average values of weighted (weight is defined relative to the adjacent angle) triangle normals, which intersect in a given vertex.

5.2.3 Collision Detection Between Simple Geometric Shapes

In the previous chapter we analyzed collisions between rays and polygons. Though objects can always be represented with polygons, it is often easier to compute collisions based on an object's geometric properties. This chapter will review some basic concepts for collision detection that are based on object geometry. In order to simplify analysis, we will mostly limit collisions to a single plane only.

First we will consider a collision between a sphere and a dimensionless particle (Fig. 5.12). Collision detection in this case is relatively simple. Based on relations in Fig. 5.12 it is evident that the particle collided with the sphere when the length of vector $\mathbf{p}_{12} = \mathbf{p}_2 - \mathbf{p}_1$ is smaller than the sphere radius r. Sphere deformation can be computed as

$$d = \begin{cases} 0 & \text{for} \quad \|\mathbf{p}_{12}\| > r \\ r - \|\mathbf{p}_{12}\| & \text{for} \quad \|\mathbf{p}_{12}\| < r. \end{cases} \tag{5.18}$$

In the case of a frictionless collision, the reaction force direction is determined along vector $\mathbf{p}_{12}$, which is normal to the sphere surface.

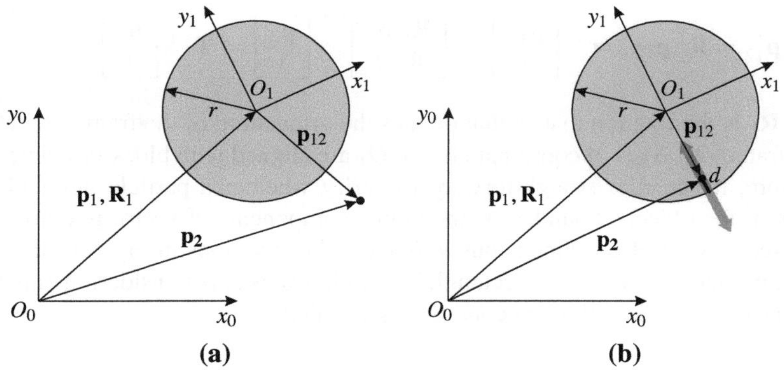

Fig. 5.12 Collision between a *sphere* and a dimensionless particle (simplified view with a collision between a *circle* and a particle); the *left* image shows relations before the collision, while the *right* image shows relations after the collision. *Thickened straight arrows* indicate force directions

Figure 5.13 shows collision between a block and a dimensionless particle. As in the case of collision with a sphere, vector $\mathbf{p}_{12} = \mathbf{p}_2 - \mathbf{p}_1$ should first be computed. However, this is not sufficient, since it is necessary to determine the particle position relative to individual block faces (sides of the rectangle on Fig. 5.13). Namely, vector $\mathbf{p}_{12}$ is computed relative to the global coordinate frame O_0, while block faces are in general not aligned with axes of frame O_0. Collision detection can be simplified by transforming vector $\mathbf{p}_{12}$ into the local coordinate frame O_1, resulting in $\mathbf{p}_{12}^1$. Vector $\mathbf{p}_{12}^1$ can be computed as

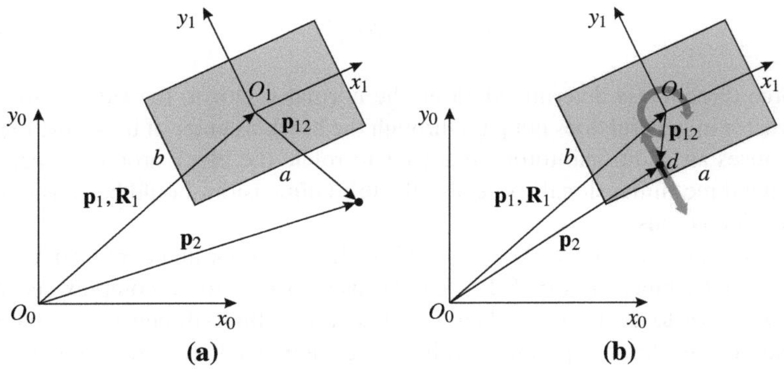

Fig. 5.13 Collision between a block and a dimensionless particle (simplified view with a collision between a *rectangle* and a particle); *left* image shows relations before the collision, while the *right* image shows relations after the collision. *Thickened straight arrows* indicate force directions, while *thickened circular arrow* indicates torque acting on the block

$$\mathbf{p}_{12}^1 = \mathbf{R}_1^T \mathbf{p}_{12} \quad \text{or} \quad \begin{bmatrix} \mathbf{p}_{12}^1 \\ 1 \end{bmatrix} = \begin{bmatrix} \mathbf{R}_1 & \mathbf{p}_1 \\ \mathbf{0} & 1 \end{bmatrix}^{-1} \begin{bmatrix} \mathbf{p}_2 \\ 1 \end{bmatrix} = \mathbf{T}^{-1} \begin{bmatrix} \mathbf{p}_2 \\ 1 \end{bmatrix}, \tag{5.19}$$

where $\mathbf{R}_1$ is the rotation matrix that defines the orientation of the frame O_1 relative to the frame O_0. Axes of coordinate frame O_1 are aligned with block principle axes. Therefore, it becomes straightforward to verify whether a particle lies within or outside of the object's boundaries. Individual components of vector $\mathbf{p}_{12}^1$ have to be compared against block dimensions a, b and c. For relations in Fig. 5.13 it is clear that the particle lies within the rectangle's boundaries (we are considering only plane relations here), if the following condition is satisfied

$$|p_{12_x}^1| < \frac{a}{2} \quad \wedge \quad |p_{12_y}^1| < \frac{b}{2}, \tag{5.20}$$

where $p_{12_x}^1$ and $p_{12_y}^1$ are x and y components of vector $\mathbf{p}_{12}^1$.

However, in this case it is not trivial to determine deformation d and reaction force direction. Namely, here we have to take into account also the relative position and direction of motion between the block and the particle at the instant before the collision occurrence. If the collision occurred along the side a (see Fig. 5.13), then the resulting deformation equals

$$d = \frac{b}{2} - |p_{12_y}^1|, \tag{5.21}$$

and force direction for frictionless contact is along the normal vector to the side a. In the opposite case

$$d = \frac{a}{2} - |p_{12_x}^1|, \tag{5.22}$$

and force direction is determined along the normal vector to the side b. Since the force vector in general does not pass through the block's center of mass, the reaction force causes an additional torque that tries to rotate the block around its center of mass. In a three-dimensional space also the third dimension should be considered in the above equations.

Transformation of a pose of one object into the local coordinate frame of the other object, as determined by Eq. (5.19), can be used also in more complex scenarios, where we have to deal with collisions between two three-dimensional objects (it can also be used for computing a collision between a sphere and a dimensionless particle). Figure 5.14 shows a collision between two spheres. Collision analysis in this case is as simple as analysis of collision between a sphere and a particle. It is only necessary to compute vector $\mathbf{p}_{12}$. If its length is smaller than the sum of sphere radii $r_1 + r_2$, the two spheres collided and the total deformation of both spheres equals

$$d = \begin{cases} 0 & \text{for} \quad \|\mathbf{p}_{12}\| > r_1 + r_2 \\ r_1 + r_2 - \|\mathbf{p}_{12}\| & \text{for} \quad \|\mathbf{p}_{12}\| < r_1 + r_2. \end{cases} \tag{5.23}$$

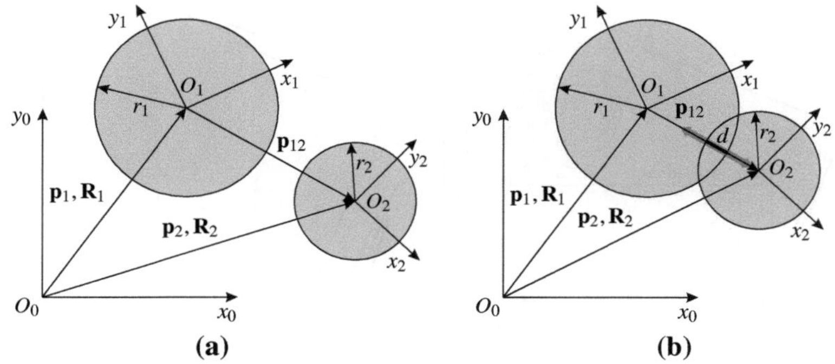

Fig. 5.14 Collision between two spheres (simplified view with a collision between two *circles*); the *left* image shows relations before the collision, while the *right* image shows relations after the collision. *Thickened straight arrows* indicate force directions

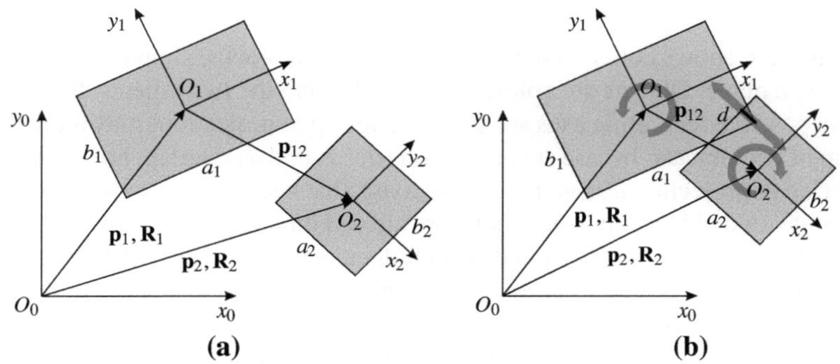

Fig. 5.15 Collision between two blocks (simplified view with a collision between two *rectangles*); the *left* image shows relations before the collision, while the *right* image shows relations after the collision. *Thickened straight arrows* indicate force directions, while *thickened circular arrows* indicates torques acting on the blocks

Deformation of an individual sphere can be computed based on the stiffness values of both objects (for example $d_1 = k_2/(k_1 + k_2)d$). In the case of frictionless collision the reaction force direction is determined along the vector $\mathbf{p}_{12}$.

Analysis of collision between two blocks is much more complex than collision detection between two spheres. Relations are shown in Fig. 5.15.

The analysis will be based on the following observation. Two convex polyhedrons are separated and do not intersect, if they can be separated by a plane parallel to one of the surfaces of the two polyhedrons or with a plane that includes one of the edges of both polyhedrons.

Existence of such a plane can be determined by projections of polyhedrons on axes that are perpendicular to the previously mentioned planes. Two convex polyhedrons are separated if there exist such an axis, on which their projections are separated. Such an axis is called a *separating axis*. If such an axis cannot be found, the two polyhedrons intersect.

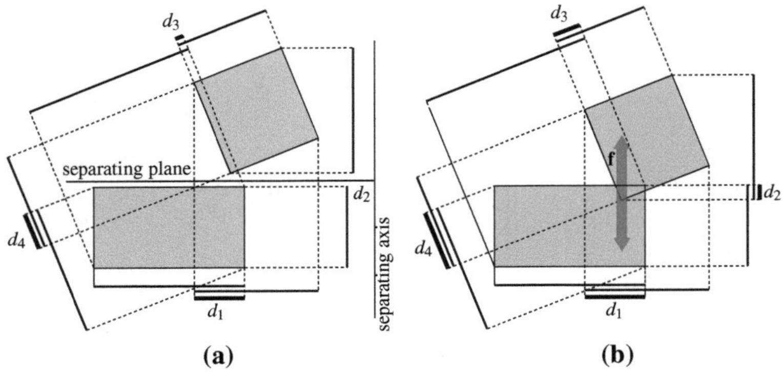

Fig. 5.16 Collision between two blocks (simplified view with a collision between two *rectangles*)—separating axis and separating plane are indicated; the *left* image shows relations before the collision, while the *right* image shows relations after the collision

Figure 5.16 shows collision detection between two blocks, simplified as two rectangles in a plane. Objects are colored in grey. Around the two objects, their projections on possible separating axes are shown. Overlap of projections may indicate also overlap of two objects: in case (a) projections d_1, d_3 and d_4 overlap, however there is no overlap of projection d_2 (on the vertical axis), that becomes the separating axis. In case (b) the two objects intersect, therefore also all their projections overlap. In case (a) a separating plane can be found, which does not exist in case (b). Therefore, it is possible to conclude that in case (b) the two blocks intersect and are in contact. Thus, collision occurred.

As an additional result of the analyzed collision detection algorithm, the overlap between the two blocks can be estimated. Once we compute the penetration of one object into the other, the reaction forces on the two objects can be computed. Force direction can be determined with the direction of the separating axis with the smallest overlap. In the case shown in Fig. 5.16, projection d_2 results in the smallest overlap. Thus force direction is aligned with the vertical separating axis parallel to d_2. Since a force vector in general does not pass through the block's center of mass, the reaction force causes an additional torque that tries to rotate the block around its center of mass.

The example shown on Fig. 5.16 shows collision detection for two rectangles in a plane. We will now consider the collision detection for blocks in a three-dimensional space [10]. Each block can be represented with its center point $\mathbf{p}_i = [p_{i_x}, p_{i_y}, p_{i_z}]^T$, its orientation is defined with a set of block principal axes $\mathbf{x}_i$, $\mathbf{y}_i$, $\mathbf{z}_i$, such that the orientation of the block is

$$\mathbf{R} = [\mathbf{x}_i, \mathbf{y}_i, \mathbf{z}_i], \tag{5.24}$$

and block dimensions are a_i, b_i, c_i. For two blocks in three-dimensional space there are 15 possible separating axes. The first three potential separating axes are directed along the principal axes of the first block $\mathbf{x}_1$, $\mathbf{y}_1$, $\mathbf{z}_1$, the next three in the direction of the principle axes of the second block $\mathbf{x}_2$, $\mathbf{y}_2$, $\mathbf{z}_2$ and the other nine possible separating axes along vectors of all combinations of cross products of principle axes of the first and second block. Let the vector $\mathbf{L}_j$ be the jth possible separating axis, where $j = 1, 2, \ldots, 15$.

Both blocks are projected on the possible separating axis $\mathbf{L}_j$ and if the two projected intervals do not intersect, the axis $\mathbf{L}_j$ is a separating axis. Two projected intervals do not intersect if the distance between the centers of the intervals is larger than the sum of radii of the intervals.

The radius of the projected interval of the ith block on axis $\mathbf{L}_j$ is

$$\frac{1}{\mathbf{L}_j \cdot \mathbf{L}_j}(a_i \mathrm{sgn}(\mathbf{L}_j \cdot \mathbf{x}_i)\mathbf{L}_j \cdot \mathbf{x}_i + b_i \mathrm{sgn}(\mathbf{L}_j \cdot \mathbf{y}_i)\mathbf{L}_j \cdot \mathbf{y}_i + c_i \mathrm{sgn}(\mathbf{L}_j \cdot \mathbf{z}_i)\mathbf{L}_j \cdot \mathbf{z}_i), \quad (5.25)$$

while the distance between the centers of intervals is the length of the vector $\mathbf{p}_{12}$ projected on axis $\mathbf{L}_j$ calculated as

$$\frac{1}{\mathbf{L}_j \cdot \mathbf{L}_j}|\mathbf{L}_j \cdot \mathbf{p}_{12}|. \quad (5.26)$$

Since both expressions (5.25) and (5.26) are divided with $\frac{1}{\mathbf{L}_j \cdot \mathbf{L}_j}$, normalized radii can be written as

$$r_{i,j} = a_i \mathrm{sgn}(\mathbf{L}_j \cdot \mathbf{x}_i)\mathbf{L}_j \cdot \mathbf{x}_i + b_i \mathrm{sgn}(\mathbf{L}_j \cdot \mathbf{y}_i)\mathbf{L}_j \cdot \mathbf{y}_i + c_i \mathrm{sgn}(\mathbf{L}_j \cdot \mathbf{z}_i)\mathbf{L}_j \cdot \mathbf{z}_i, \quad (5.27)$$

and normalized distance between centers of intervals can be written as

$$r_{12,j} = |\mathbf{L}_j \cdot \mathbf{p}_{12}|. \quad (5.28)$$

The non-intersection test for the jth possible separating axis is then

$$r_{12,j} > r_{1,j} + r_{2,j}. \quad (5.29)$$

To compute the test for each possible axis, a rotation matrix $\mathbf{R}_{12}$ between the first and the second block is calculated as

$$\mathbf{R}_{12} = \mathbf{R}_1^T \mathbf{R}_2. \quad (5.30)$$

Components of the matrix $\mathbf{R}_{12}$ are c_{ij} such that the matrix can be written as

$$\mathbf{R}_{12} = \begin{bmatrix} c_{00} & c_{01} & c_{02} \\ c_{10} & c_{11} & c_{12} \\ c_{20} & c_{21} & c_{22} \end{bmatrix}. \quad (5.31)$$

Table 5.1 gives the 15 non-intersection tests for each of the 15 possible separating axes. To test if the blocks are in collision the non-intersection test is done for each of

Table 5.1 Table gives values for r_{12}, r_1, r_1 for the non-intersection test $r_{12} > r_1 + r_2$

L	r_1	r_1	r_{12}
$\mathbf{x}_1$	a_1	$a_2\lvert c_{00}\rvert + b_2\lvert c_{01}\rvert + c_2\lvert c_{02}\rvert$	$\lvert \mathbf{x}_1 \cdot \mathbf{p}_{12}\rvert$
$\mathbf{y}_1$	b_1	$a_2\lvert c_{10}\rvert + b_2\lvert c_{11}\rvert + c_2\lvert c_{12}\rvert$	$\lvert \mathbf{y}_1 \cdot \mathbf{p}_{12}\rvert$
$\mathbf{z}_1$	c_1	$a_2\lvert c_{20}\rvert + b_2\lvert c_{21}\rvert + c_2\lvert c_{22}\rvert$	$\lvert \mathbf{z}_1 \cdot \mathbf{p}_{12}\rvert$
$\mathbf{x}_2$	$a_1\lvert c_{00}\rvert + b_1\lvert c_{10}\rvert + c_1\lvert c_{20}\rvert$	a_2	$\lvert \mathbf{x}_2 \cdot \mathbf{p}_{12}\rvert$
$\mathbf{y}_2$	$a_1\lvert c_{01}\rvert + b_1\lvert c_{11}\rvert + c_1\lvert c_{21}\rvert$	b_2	$\lvert \mathbf{y}_2 \cdot \mathbf{p}_{12}\rvert$
$\mathbf{z}_2$	$a_1\lvert c_{02}\rvert + b_1\lvert c_{12}\rvert + c_1\lvert c_{22}\rvert$	c_2	$\lvert \mathbf{z}_2 \cdot \mathbf{p}_{12}\rvert$
$\mathbf{x}_1 \times \mathbf{x}_2$	$b_1\lvert c_{20}\rvert + c_1\lvert c_{10}\rvert$	$b_2\lvert c_{02}\rvert + c_2\lvert c_{01}\rvert$	$\lvert c_{10}\mathbf{z}_1 \cdot \mathbf{p}_{12} - c_{20}\mathbf{y}_1 \cdot \mathbf{p}_{12}\rvert$
$\mathbf{x}_1 \times \mathbf{y}_2$	$b_1\lvert c_{21}\rvert + c_1\lvert c_{11}\rvert$	$a_2\lvert c_{02}\rvert + c_2\lvert c_{00}\rvert$	$\lvert c_{11}\mathbf{z}_1 \cdot \mathbf{p}_{12} - c_{21}\mathbf{y}_1 \cdot \mathbf{p}_{12}\rvert$
$\mathbf{x}_1 \times \mathbf{z}_2$	$b_1\lvert c_{22}\rvert + c_1\lvert c_{12}\rvert$	$a_2\lvert c_{01}\rvert + b_2\lvert c_{00}\rvert$	$\lvert c_{12}\mathbf{z}_1 \cdot \mathbf{p}_{12} - c_{22}\mathbf{y}_1 \cdot \mathbf{p}_{12}\rvert$
$\mathbf{y}_1 \times \mathbf{x}_2$	$a_1\lvert c_{20}\rvert + c_1\lvert c_{00}\rvert$	$b_2\lvert c_{12}\rvert + c_2\lvert c_{11}\rvert$	$\lvert c_{20}\mathbf{x}_1 \cdot \mathbf{p}_{12} - c_{00}\mathbf{z}_1 \cdot \mathbf{p}_{12}\rvert$
$\mathbf{y}_1 \times \mathbf{y}_2$	$a_1\lvert c_{21}\rvert + c_1\lvert c_{01}\rvert$	$a_2\lvert c_{12}\rvert + c_2\lvert c_{10}\rvert$	$\lvert c_{21}\mathbf{x}_1 \cdot \mathbf{p}_{12} - c_{01}\mathbf{z}_1 \cdot \mathbf{p}_{12}\rvert$
$\mathbf{y}_1 \times \mathbf{z}_2$	$a_1\lvert c_{22}\rvert + c_1\lvert c_{02}\rvert$	$a_2\lvert c_{11}\rvert + c_2\lvert c_{10}\rvert$	$\lvert c_{22}\mathbf{x}_1 \cdot \mathbf{p}_{12} - c_{02}\mathbf{z}_1 \cdot \mathbf{p}_{12}\rvert$
$\mathbf{z}_1 \times \mathbf{x}_2$	$a_1\lvert c_{10}\rvert + b_1\lvert c_{00}\rvert$	$b_2\lvert c_{22}\rvert + c_2\lvert c_{21}\rvert$	$\lvert c_{00}\mathbf{y}_1 \cdot \mathbf{p}_{12} - c_{10}\mathbf{x}_1 \cdot \mathbf{p}_{12}\rvert$
$\mathbf{z}_1 \times \mathbf{y}_2$	$a_1\lvert c_{11}\rvert + b_1\lvert c_{01}\rvert$	$a_2\lvert c_{22}\rvert + c_2\lvert c_{20}\rvert$	$\lvert c_{01}\mathbf{y}_1 \cdot \mathbf{p}_{12} - c_{11}\mathbf{x}_1 \cdot \mathbf{p}_{12}\rvert$
$\mathbf{z}_1 \times \mathbf{z}_2$	$a_1\lvert c_{12}\rvert + b_1\lvert c_{02}\rvert$	$a_2\lvert c_{21}\rvert + b_2\lvert c_{20}\rvert$	$\lvert c_{02}\mathbf{y}_1 \cdot \mathbf{p}_{12} - c_{12}\mathbf{x}_1 \cdot \mathbf{p}_{12}\rvert$

the 15 possible separating axes. If separating axis is found, the blocks do not collide and the testing of intersection is stopped. If none of the test is passed the blocks are in collision.

Figures 5.17 and 5.18 show relations during a collision between a block and a sphere. Also in this case it is possible to compute collisions between the two objects based on the knowledge obtained in the previous paragraphs. As in the case of collisions between two blocks, it is necessary to compute separating planes. If a separating plane does not exist, the two objects intersect. The separating axis with

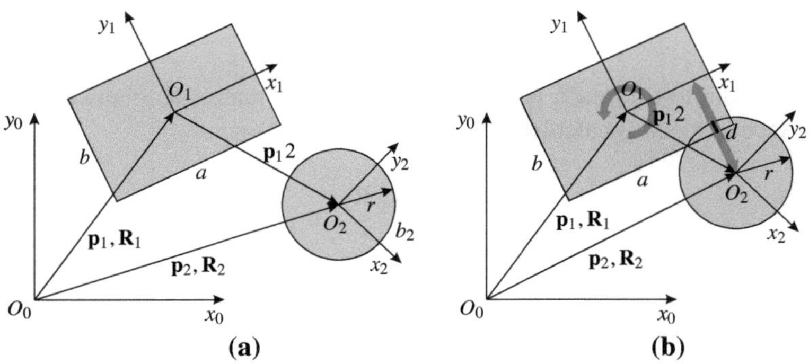

Fig. 5.17 Collision between a block and a sphere (simplified view with a collision between a *rectangle* and a *circle*); *left* image shows relations before the collision, while the *right* image shows relations after the collision. *Thickened straight arrows* indicate force directions, while *thickened circular arrow* indicates torque acting on the block

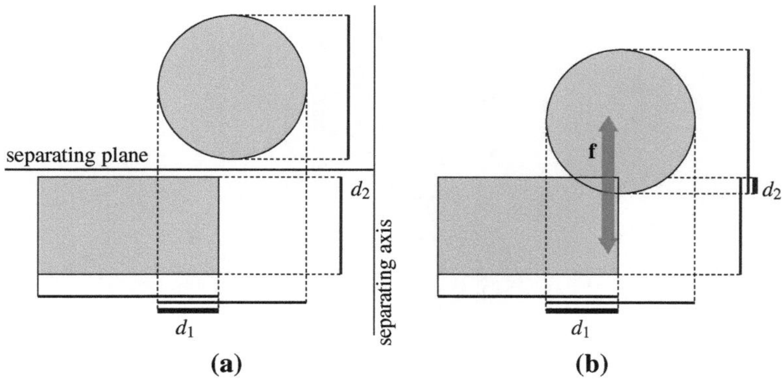

Fig. 5.18 Collision between a block and a *sphere* (simplified view with a collision between a *rectangle* and a *circle*)—separating axis and separating plane are indicated; *left* image shows relations before the collision, while the *right* image shows relations after the collision

the smallest overlap can serve as an estimation of reaction force direction, while the overlap itself determines the reaction force amplitude.

Finally, we have to consider also the problem of collisions between more complex objects. The collision detection in such a case becomes computationally more demanding. However, it can be simplified by the use of bounding volumes as shown in Fig. 5.19. The method requires that the object being involved in collision detection is embedded into the smallest possible bounding volume. The bounding volume can take the shape of a sphere, a block or a more complex geometry, such as a capsule. If a sphere is used, the bounding volume is called a *BS—bounding sphere*. The sphere is the simplest geometry of a bounding volume and it enables the easiest collision detection that does not take into account the object orientation.

The use of a bounding box enables two different approaches. Method *AABB—axis aligned bounding box* assumes that box axes are always aligned with the axes

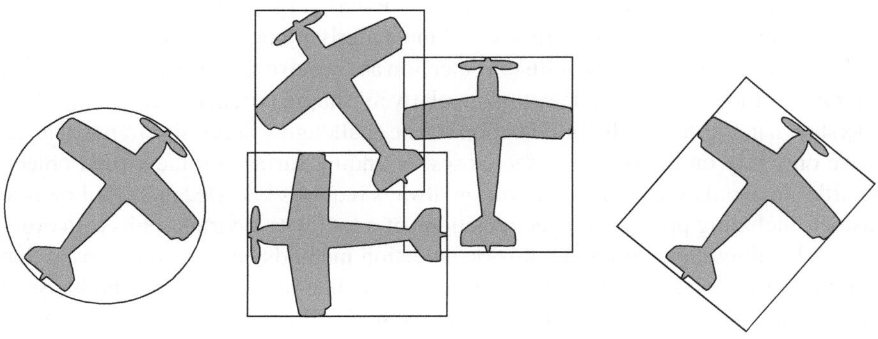

Fig. 5.19 Simplification of collision detection between complex objects with the use of bounding volumes

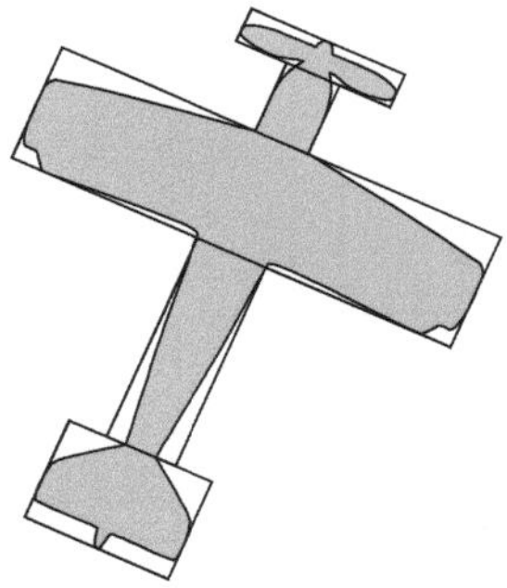

Fig. 5.20 Simplification of collision detection using multiple oriented *bounding boxes*

of the global coordinate frame, regardless of the actual object orientation. Therefore, it becomes necessary to adjust the size of the bounding box during rotation of the object in space (middle view on Fig. 5.19). Method *OBB—oriented bounding box* assumes that the object is embedded into the smallest possible bounding box that rotates together with the rotation of the object. In this case the bounding volume does not need to be adjusted during the rotation of the object. At the same time, *OBB* usually guarantees the most optimal representation of an object with a simplified bounding volume.

During the computation of collisions between objects the original (complex) geometry is replaced by a simplified geometry defined by a bounding volume. Collision detection between complex shapes can thus be translated into one of the methods addressed previously in this chapter.

The use of bounding volumes for collision detection allows only approximation of true collisions between objects. If a simple bounding volume does not give satisfactory results, the object can be split into smaller parts and each of these parts can be embedded into its own bounding volume. The use of multiple oriented bounding boxes for a representation of a virtual object is shown in Fig. 5.20. Such representation enables a more detailed approximation of the underlying object geometry.

The analyzed methods for collision detection are some of the simplest methods and guarantee efficient collision detection in relatively simple virtual environments. As it was shown in Fig. 5.7, collision detection can be challenging in certain circumstances, if we only rely on the search of the closest bounding surface on the virtual object. In order to avoid this, collision detection always requires knowledge of tool motion history, including previous contact positions of a tool with a virtual object. In order to avoid ambiguities, different collision detection methods have been proposed that eliminate weaknesses of the presented simple collision detection methods. Such methods are the so-called *godfather-object* method [11] and the method with a *proxy object* [12].

References

1. Barraff, D.: Fast contact force computation for nonpenetrating rigid bodies. In: Computer Graphics (SIGGRAPH Proceedings), pp. 23–34. Orlando (1994)
2. Gottschalk, S.: Collision detection techniques for 3D models. Cps 243 term paper, University of North Carolina (1997)
3. Lin, M., Gottschalk, S.: Collision detection between geometric models: a survey. In: Proceedings of IMA Conference on Mathematics on Surfaces, pp. 11–19 (1998)
4. Adachi, Y., Kumano, T., Ogino, K.: Intermediate representation for stiff virtual objects. In: Proceedings of the Virtual Reality Annual International Symposium, pp. 203–210 (1995)
5. Konig, H., Strohotte, T.: Fast collision detection for haptic displays using polygonal models. In: Proceedings of the Conference on Simulation and Visualization, pp. 289–300. Ghent (2002)
6. Moller, T., Trumbore, B.: Fast, minimum storage ray/triangle intersection. J. Graph. Tools **2**, 21–28 (1997)
7. Segura, R.J., Feito, F.R.: Algorithms to test ray-triangle intersection. Comparative study. J. WSCG **9** (2001)
8. Badouel, D.: An efficient ray-polygon intersection. In: Glassner, S.A. (ed.) Graphics Gems. Academic Press Inc, London (1990)
9. Basdogan, C., Ho, C.H., Srinivasan, M.A.: A ray-based haptic rendering technique for displaying shape and texture of 3D objects in virtual environments. In: Proceedings of the ASME Dynamics Systems and Control Division, pp. 77–84 (1997)
10. Eberly, D.: Dynamic collision detection using oriented bounding boxes. Technical Report, Magic Software Inc (2007)
11. Zilles, C., Salisbury, J.K.: A constraint-based god-object for haptic display. In: Proceedings of the International Conference on Intelligent Robots and Systems, pp. 146–151 (1995)
12. Ruspini, D.C., Koralov, K., Khatib, O.: The haptic display of complex graphical environments. In: Computer Graphics (SIGGRAPH Proceedings), pp. 345–352. Los Angeles, California (1997)

References

Chapter 6
Haptic Rendering

Computer haptics is a research area dealing with techniques and processes related to generation and rendering of contact properties in a virtual environment and displaying of this information to a human user via a haptic interface. Computer haptics deals with models and properties of virtual objects, as well as algorithms for displaying haptic feedback in real time.

Haptic interfaces provide haptic feedback information about the computer generated or remote environment to a user who interacts with this environment. Since these interfaces do not have their own intelligence, they only allow presentation of computer generated quantities. For this purpose it is necessary to understand physical models of a virtual environment that enable generation of time dependent variables (forces, accelerations, vibrations, temperature, ...) required for control of the haptic interface.

The physical model defines dynamic properties of virtual objects in a simulated environment that is usually based on Newtonian physics. Various physical model aspects need to be taken into account while planning haptic interaction in a virtual environment. For example, object deformations (plastic or elastic) are the result of forces and torques applied on the object. On the other hand, object textures and physical constraints, such as gravity and friction, contribute to a realistic haptic interaction.

The task of *haptic rendering* is to enable the user to touch, sense and manipulate virtual (real) objects in a simulated (remote) environment via a haptic interface [1, 2]. The basic idea of haptic rendering can be explained using Fig. 6.1, where a frictionless sphere is positioned in the origin of a virtual environment. Now assume that the user interacts with a sphere in a single point, which is defined by the haptic interface end-effector position (**HIP**) [3]. In a real world this would be analogous to touching a sphere with a tip of a thin stick. When moving through a free space the haptic interface behaves *passively* and does not apply any force onto the user until the occurrence of a contact with a sphere. Since the sphere has final stiffness, the **HIP** penetrates into the sphere at the point of contact. When the contact with the sphere is detected, the corresponding reaction force is computed and transmitted via a haptic interface to the user. The haptic interface becomes *active* and generates a reaction force that prevents further penetration into the object. The magnitude of the reaction

M. Mihelj and J. Podobnik, *Haptics for Virtual Reality and Teleoperation*, Intelligent Systems, Control and Automation: Science and Engineering 64,
DOI: 10.1007/978-94-007-5718-9_6,

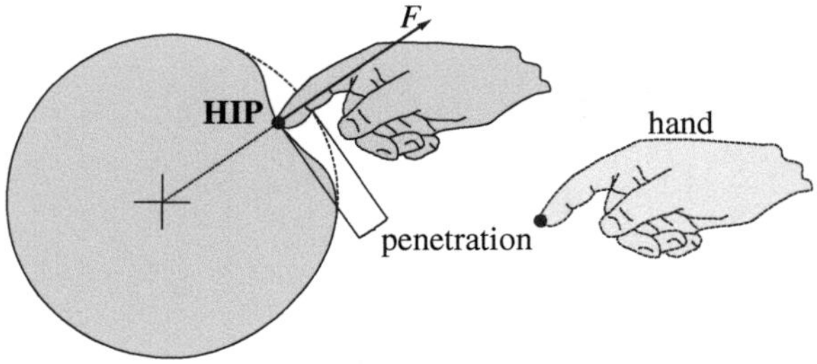

Fig. 6.1 Haptic rendering of a sphere in a virtual environment

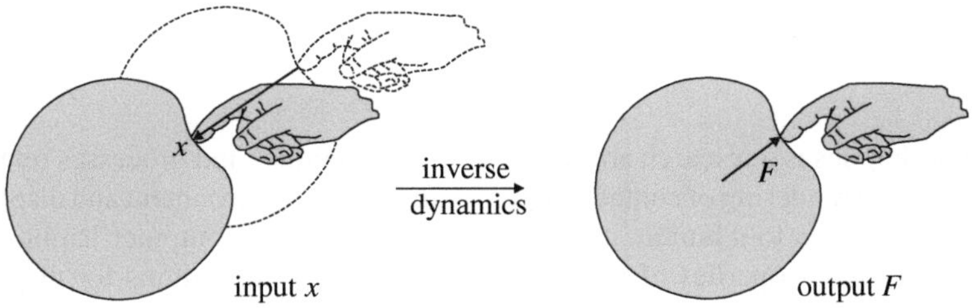

Fig. 6.2 Stiffness (impedance) model of haptic interaction

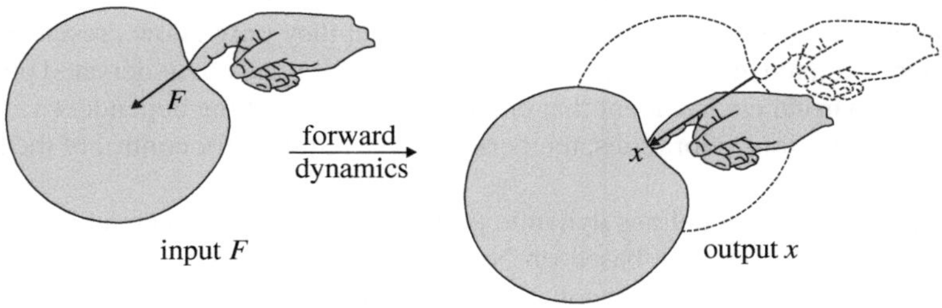

Fig. 6.3 Compliance (admittance) model of haptic interaction

force can be computed based on a simple assumption that the force is proportional to the penetration depth. With the assumption of a sphere without friction, the reaction force direction is determined with a vector normal to the sphere surface at the point of contact.

In general we distinguish two models of haptic interaction with the environment. The first model is called a *compliance model*, while the second is called a *stiffness model*. The two terms refer to a simple elastic model $F = Kx$ of a wall with a stiffness K, penetration depth x and reaction force F. The two concepts for modeling haptic interaction with the environment are shown in Figs. 6.2 and 6.3.

- In the case of a stiffness model in Fig. 6.2, the haptic interface measures displacement x and the simulation returns the corresponding force F as a result of relation $F = Kx$. Haptic interfaces that are excellent force sources are suitable for implementation of such a model.

- In the case of compliance model shown in Fig. 6.3, the haptic interface measures force F between the user and the haptic display and the simulation returns displacement x as a result of relation $x = K^{-1}F = CF$, where compliance C is defined as the inverse value of stiffness K. Stiff haptic interfaces, such as industrial manipulators equipped with a force and torque sensor, are suitable for implementation of such a model.

In the case of more complex models, where in addition to compliance also viscous and inertial components are present, the terms stiffness model and compliance model are substituted with terms *impedance model* as an equivalent of the stiffness model and *admittance model* as an equivalent of the compliance model. For the purpose of generality we will use terms impedance and admittance in the following sections. When an object is displaced due to a contact, object dynamics needs to be considered to determine the relation between force and displacement. An inverse dynamic model is required for computing the impedance and a forward dynamic model is required for computing the admittance causality structure.

Most approaches to haptic rendering are based on the assumption of interactions with stiff grounded objects, where stiffness characteristics dominate over other dynamic properties. In the case of objects in a virtual or remote environment that are displaced as a result of haptic interactions, the object deformation is masked with the object displacement. In such cases it is reasonable to attribute the entire haptic interface displacement to the object movement without considering the object deformation. Namely, we can assume that the majority of real objects do not deform considerably under the contact forces. However, such assumption is not valid when the impedance, which is the result of object displacement, is approximately equal to the impedance properties for object deformation.

6.1 Modeling of Free Space

A system for haptic rendering must be capable of realistically rendering motion in a free space. In order to make this possible, a haptic interface must possess characteristics that enable its movement without significant physical effort and with a smallest possible disturbance resulting from friction in the mechanism, intrinsic inertia and vibrations. This can be achieved passively with the mechanism design that is based on a small intrinsic impedance (small inertia, damping and elasticity), or actively, where the control system compensates intrinsic impedance of the device.

6.2 Modeling of Object Stiffness

A critical test of a haptic display and control algorithm represents the maximal stiffness of objects that can be realistically rendered and displayed with a haptic interface. Haptic interface bandwidth, stiffness, resolution and the capability of generating

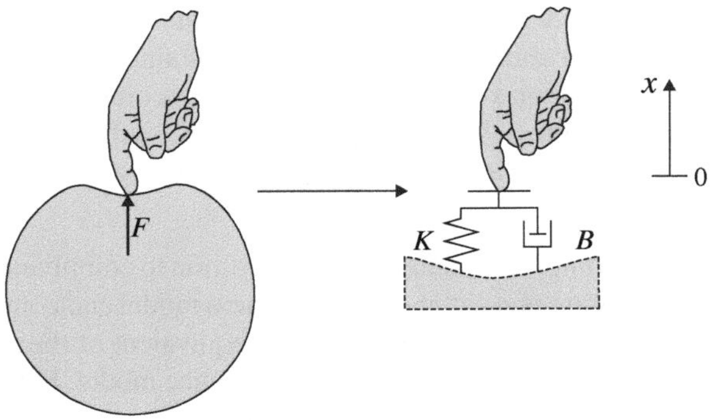

Fig. 6.4 Spring-directed damper model of a contact with a grounded object

forces are quantities that decisively affect accuracy of displaying transient responses during contacts with virtual objects. In addition, the contact force modeling approach affects contact intensity and system stability.

At the moment of collision with the surface of a virtual object, an instantaneous change of force occurs with an impulse large enough to withdraw the momentum from the user's hand or a tool. This requires high frequency bandwidth that enables fast and stable change of force. On the other hand, continuous contact with the object surface requires large haptic interface end-effector forces without saturation of device actuators and unstable behavior. Accurate transient responses and stable steady state behavior enable realistic impression of object stiffness.

The most common way of modeling stiff and grounded surface is based on a model consisting of a parallel connection of a spring with stiffness K and a damper with viscosity B, as shown in Fig. 6.4. Given $x = 0$ for an undeformed object and $x < 0$ inside the object boundaries, the modeled contact force on the virtual hand as shown in Fig. 6.4 equals

$$F = \begin{cases} -(Kx + B\dot{x}) & \text{for} \quad x < 0 \wedge \dot{x} < 0 \\ -Kx & \text{for} \quad x < 0 \wedge \dot{x} \geq 0 \\ 0 & \text{for} \quad x \geq 0. \end{cases} \tag{6.1}$$

In this case, the viscous damping behaves as a directed damper, that is active during the penetration into the object and passive during the withdrawal from the object. This enables a stable and damped contact with the object as well as a realistic contact rendering. Contact relations are shown in Fig. 6.5 for a one degree of freedom system. From relations shown in Fig. 6.5 it is apparent that at the instance of the contact, there is a step change in the force signal due to the contribution of viscous damping, since the approach velocity differs from zero. During the penetration into the object the influence of the viscous part is being reduced due to the decreasing movement velocity. At the same time the contribution of the elastic element increases due to the increased depth of penetration into the object. At the instance of the largest object deformation, the penetration velocity equals zero and the reaction force is

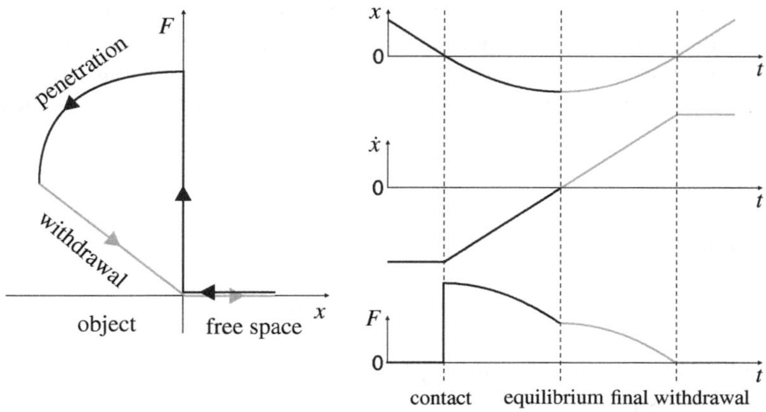

Fig. 6.5 Relations during a contact simulation with a spring-directed damper model

only the result of the elastic element. Since the damper operates in a single direction (only active during penetration and inactive during withdrawal), this results in a linearly decreasing reaction force as a function of displacement x. The reaction force reaches zero value at the boundary of the undeformed object, resulting in a smooth transition between the object and a free space. Such a modeling approach guarantees pronounced initial contact, rigidity of a stiff surface and a smooth withdrawal from the object surface.

The spring-damper model can also be analyzed from the perspective of human perception. Namely, the essence of haptic rendering is to convince the user in the realism of contact with a virtual object. Thus, human perception capabilities need to be considered as well. *Perceived stiffness* [4] is based on psychophysiological factors that determine the rigidity of the contact. Perceived stiffness H is defined as a ratio of initial rate of force change $\dot{F}$ and movement velocity $\dot{x}$ at the instance of initial contact

$$H = \frac{\dot{F}}{\dot{x}}. \tag{6.2}$$

The perceived stiffness emphasizes how fast the change of force is in relation to the movement velocity during the transient response at the initial contact. A surface stiffness in a steady state is less important, since a human's capability of discrimination between different stiffness values is rather limited [5].

The spring-damper model (since we are only interested in the initial contact, we are neglecting the directional nature of the damper) can be written in a Laplace form as

$$F(s) = (Bs + K)x(s), \tag{6.3}$$

where $x = 0$ indicates an undeformed object surface. Since a haptic interface hits the object with a velocity $\dot{x}$, this means a nonlinear increase in force F. Since a haptic interface behaves as a low-pass filter due to the limited bandwidth of the device actuators, finite intrinsic admittance and delays in the control loop, it is justified to introduce a low-pass filter $a/(s+a)$, which approximates the dynamics of the haptic interface, into the contact model. The entire contact model can now be written as

$$F(s) = \frac{a}{s+a}(Bs + K)x(s) \tag{6.4}$$

and in a time space using the inverse Laplace transform as

$$\dot{F} = -aF + aB\dot{x} + aKx. \tag{6.5}$$

At the moment of initial contact, the position equals $x = 0$ and the force equals $F = 0$ and consequently

$$\dot{F} = aB\dot{x} \quad \Longrightarrow \quad H = \frac{\dot{F}}{\dot{x}} = aB. \tag{6.6}$$

With the added viscous damping the perceived stiffness equals the product of a low-pass filter parameter a and viscous damping B. Since a is a design parameter of a haptic interface and cannot be easily changed, the perceived stiffness can be increased by increasing the viscous damping B. The importance of viscous damping is contradictory to the hypothesis that the increased virtual object stiffness requires an increase of the stiffness of the elastic element. However, this becomes clearer if relations in Fig. 6.5 are considered. The elastic element is primarily important during the steady state, when the movement velocity equals zero. On the other hand, during the initial contact, the viscous damping parameter defines the increase of force value during the initial contact (perceived stiffness is mainly related to this initial contact).

6.3 Friction Model

In a real world we rarely deal with frictionless surfaces. During manipulation of objects we in fact rely on friction. Therefore, in a haptic system we must consider also the friction force F_f in addition to the reaction forces resulting from object deformation.

Many models have been proposed to mathematically describe friction, from classical static models to dynamic friction models [6–8]. Richard [9] used a version of the Dahl model [10] and Karnopp model to identify friction of aluminum block sliding on hard rubber, Teflon and brass and also to display friction in one dimension. The author suggested the Karnopp model rather than the Dahl model for haptic display, because it allows the simplest entree into haptic friction identification and accurately captures sticking. Jeon et al. [11] also used the Dahl model to change the friction properties between tooltip and surface.

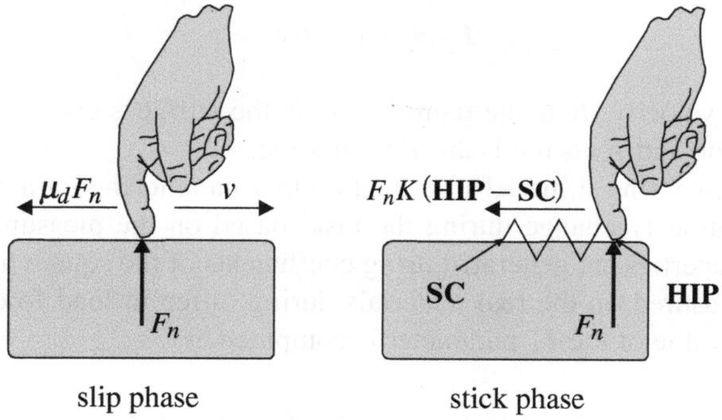

Fig. 6.6 Stick-slip friction model

Rendering of friction is often based on a stick-slip model [12] shown in Fig. 6.6, where the velocity threshold v_{min} determines the sticking region.

- The slip phase occurs when the tangential velocity relative to the object surface exceeds the predefined minimal velocity $v > v_{min}$. Friction can be modeled either using the viscous model, where friction force is proportional to the velocity, $F_f = Bv$ (B represents viscous damping), or with a Coulomb model, where friction force is proportional to the normal force F_n acting on the object surface, $F_f = \mu_d F_n$ (μ_d is the coefficient of dynamic friction).
- The sticking phase occurs when the tangential velocity relative to the object surface is smaller than the minimal threshold velocity $v \leq v_{min}$. In this case a virtual spring is attached to the object in the point **SC**. Stiffness of the virtual spring depends on the predefined stiffness K and is proportional to the normal force on the object surface F_n. When the spring is extended to the point **HIP**, it generates an opposing force $F_n K(\mathbf{HIP} - \mathbf{SC})$.

The complete model of haptic friction rendering can be summarized as:

$$F_f = \begin{cases} \mu_d F_n \ \text{ or } \ Bv & \text{during slip phase} \\ F_n K(\mathbf{HIP} - \mathbf{SC}) & \text{during stick phase.} \end{cases} \tag{6.7}$$

Such a friction model corresponds to the friction characteristics between a finger and a surface.

In the next section a modified version of the Dahl friction model with the extension of damping will be presented. This model has two main benefits. The first is that the model is simple and easy to develop. The second and also the main reason is that the model includes three parameters that are physically explainable and can therefore present a variety of materials. The model has the form

$$\frac{dF_d}{dt} = \sigma(1 - \frac{F_d}{F_c}\mathrm{sgn}(v))v, \tag{6.8}$$

$$F_f = F_d + Bv, \tag{6.9}$$

where v is the velocity, B is the damping, σ is the stiffness coefficient, F_c is the Coulomb friction and F_d is the Dahl friction force.

When implementing a model in a haptic environment, the parameters of each material continuously change during the task, based on the measured load force. The surface properties are generated using coefficients of the square fitting function that can be measured on the real materials during different load force conditions. Therefore, the value of the F_c parameter is computed as

$$F_c(F_n) = a_{w1,F_c} + a_{w2,F_c} F_n + a_{w3,F_c} F_n^2. \tag{6.10}$$

F_n is the normal direction load force and a_{wk,F_c} is the kth coefficient for the F_c parameter. The same approach is used to compute σ and B parameters.

The haptic friction model presented in (6.8) and (6.9) should meet certain limitations when rendering friction. These limitations are the issue of the unnatural physical values of the model parameters or measures to ensure stability of the model. In the current case the limitations are $\sigma > 0$, $F_c > 0$, $B > 0$. To ensure stability the B parameter might also be limited with the maximum value of damping due to steps in velocity resolution of the system. The Dahl friction model could also drift when a small bias force is present with small oscillations of position [13].

6.4 Dynamics of Virtual Environments

Until now we have limited our modeling of contact forces to interactions with grounded objects that do not move. However, in general we are interested in dynamic environments, where objects move in response to the contact. Simulation of virtual environment dynamics is an important part of the application regardless of being the dynamics of stiff objects, dynamics of deformations or dynamics of liquids.

Dynamics of stiff objects represent an important part of research in robotics and virtual reality and is not specific to haptic systems. Therefore, we will only discuss it briefly here.

Simulation of haptic interaction with three-dimensional objects that can translate and rotate represents an important task of which complexity depends on the number of objects and on the likelihood of collisions between them as well as with a virtual tool. Equations of motion must be solved and the new object pose computed in each computational step of the haptic control loop [14–16].

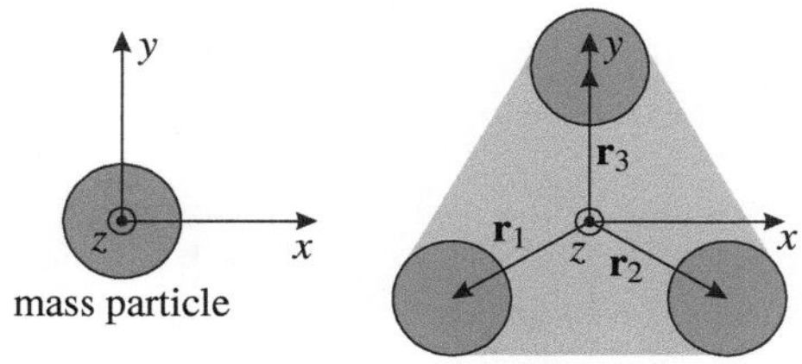

Fig. 6.7 Concept of a mass particle and a rigid body

6.4.1 Equations of Motion

Figure 6.7 shows a mass particle, which is considered a dimensionless object and a body constructed from at least three interconnected mass particles. For the purpose of further consideration we will assume that the body is composed of exactly three mass particles. The concept can be expanded for more complex objects. The coordinate system of the body is located in the object center of mass. Vectors $\mathbf{r}_i$ determine the position of each mass particle relative to the object's coordinate frame.

Although both, a mass particle as well as a body, in general move in a three-dimensional space, we will assume movement constrained to the plane for the purposes of explanation.

First we will consider the motion of a mass particle. Since we are not interested in particle orientation, the particle motion can be described using the position vector $\mathbf{p}(t)$ and its time derivatives. Velocity of a mass particle is defined as a time derivative of the position vector $\mathbf{p}(t)$ as

$$\mathbf{v}(t) = \dot{\mathbf{p}}(t) = \frac{d\mathbf{p}(t)}{dt}. \tag{6.11}$$

Knowing the motion velocity of a mass particle, its position can be computed as a time integral

$$\mathbf{p}(t) = \mathbf{p}_0 + \int_0^t \mathbf{v}(\xi)d\xi, \tag{6.12}$$

where $\mathbf{p}_0$ indicates the initial position.

The two equations above are valid also for a rigid object. However, in this case we need to consider also the object orientation. Their relations are shown in Fig. 6.8. Vector $\mathbf{p}$ and matrix $\mathbf{R}$ determine the position and orientation of the rigid object relative to the coordinate frame O_0. Object orientation can also be written in terms of quaternion q.

The position of a point $\mathbf{p}_1(t)$ located on the rigid object can be computed relative to the coordinate frame O_0 as

$$\mathbf{p}_1(t) = \mathbf{p}(t) + \mathbf{R}(t)\mathbf{r}_1, \tag{6.13}$$

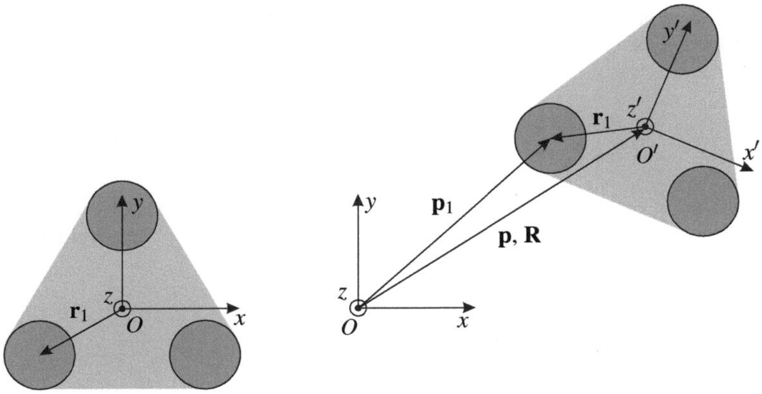

Fig. 6.8 Displacement of a rigid object from an initial pose (*left*) to a final pose (*right*)

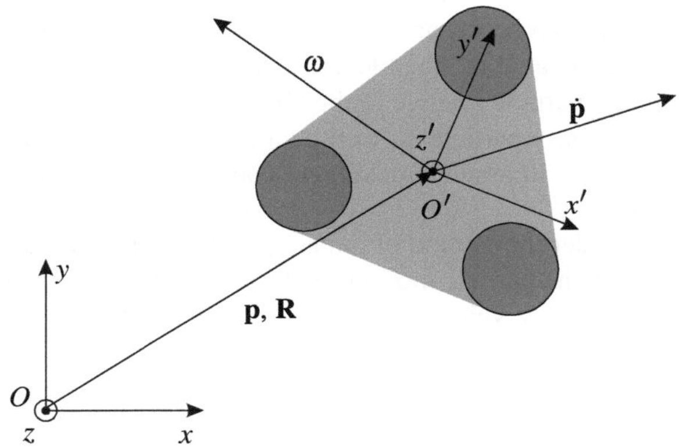

Fig. 6.9 Motion of a rigid object

or

$$\begin{bmatrix} \mathbf{p}_1(t) \\ 1 \end{bmatrix} = \begin{bmatrix} \mathbf{R}(t) & \mathbf{p}(t) \\ \mathbf{0} & 1 \end{bmatrix} \begin{bmatrix} \mathbf{r}_1 \\ 1 \end{bmatrix} = \mathbf{T}(t) \begin{bmatrix} \mathbf{r}_1 \\ 1 \end{bmatrix}, \tag{6.14}$$

where $\mathbf{T}(t)$ represents a homogenous transformation matrix.

During the motion of a rigid object we need to consider translational as well as rotational velocities and accelerations. Figure 6.9 shows the two velocity vectors.

The translational velocity is determined as a time derivative defined in Eq. (6.11) of the position vector that determines the object's center of mass. The change of orientation leads to

$$\dot{\mathbf{R}}(t) = \boldsymbol{\omega}^*(t)\mathbf{R}(t), \tag{6.15}$$

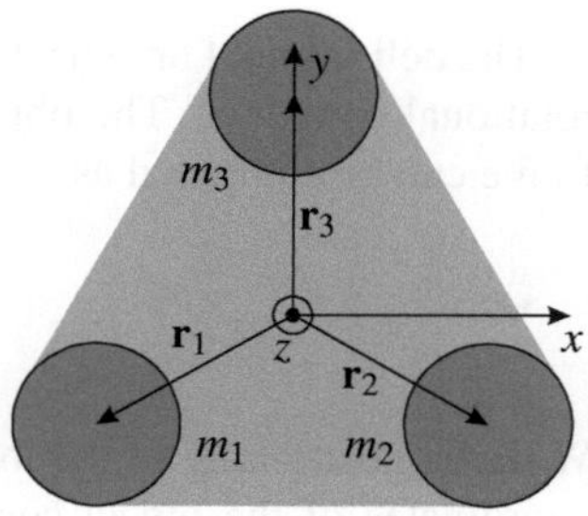

Fig. 6.10 Mass, center of mass and moment of inertia of a rigid object

when object orientation is determined in terms of a rotation matrix or

$$\dot{q}(t) = \frac{1}{2}\tilde{\omega}(t) \otimes q(t), \tag{6.16}$$

when object orientation is determined in terms of a quaternion. Matrix $\boldsymbol{\omega}^*(t)$ is a skew-symmetric matrix of vector $\boldsymbol{\omega}(t) = \left[\omega_x(t)\ \omega_y(t)\ \omega_z(t)\right]^T$, which determines the object angular velocity, such that

$$\boldsymbol{\omega}^*(t) = \begin{bmatrix} 0 & -\omega_z(t) & \omega_y(t) \\ \omega_z(t) & 0 & \omega_x(t) \\ -\omega_y(t) & \omega_x(t) & 0 \end{bmatrix}. \tag{6.17}$$

Quaternion $\tilde{\omega}(t)$ is an augmented angular velocity vector $\tilde{\omega}(t) = [0\ \ \omega_x(t)\ \ \omega_y(t)\ \ \omega_z(t)]^T$ and operator $\otimes$ denotes quaternion multiplication.

Object orientation in space can be computed with time integration of Eq. (6.15) or (6.16).

6.4.2 Mass, Center of Mass and Moment of Inertia

Object dynamic properties depend on its mass and inertia. Object mass is defined as a sum of masses of individual particles constituting the object (Fig. 6.10)

$$M = \sum_{i=1}^{N} m_i, \tag{6.18}$$

where N is the number of mass particles (in our case $N = 3$). Since objects are usually composed of homogenously distributed matter and not discrete particles, the sum in the above equation should be replaced with the integral across the object volume.

The definition of an object center of mass enables us to separate translational and rotational dynamics. The object center of mass determined in the local coordinate frame can be computed as

$$\mathbf{r}_c = \frac{\sum_{i=1}^{N} m_i \mathbf{r}_i}{M}. \tag{6.19}$$

With the positioning of the local coordinate frame in the object center of mass, the coordinates of the object center of mass expressed in the local coordinate frame equal zero. The object center of mass expressed in the global coordinate frame can be computed based on the Eq. (6.13).

Finally we define the object inertia tensor $\mathbf{I}_0$ with respect to the local coordinate frame. An object inertia tensor provides information about the distribution of object mass relative to the object center of mass

$$\mathbf{I}_0 = \sum_{i}^{N} \begin{bmatrix} m_i(r_{i_y}^2 + r_{i_z}^2) & -m_i r_{i_x} r_{i_y} & -m_i r_{i_x} r_{i_z} \\ -m_i r_{i_x} r_{i_y} & m_i(r_{i_x}^2 + r_{i_z}^2) & -m_i r_{i_y} r_{i_z} \\ -m_i r_{i_x} r_{i_z} & -m_i r_{i_y} r_{i_z} & m_i(r_{i_x}^2 + r_{i_y}^2) \end{bmatrix}, \tag{6.20}$$

where $\mathbf{r}_i = \left[r_{i_x}\ r_{i_y}\ r_{i_z}\right]^T$. If the object's shape does not change, the inertia tensor $\mathbf{I}_0$ is constant. The inertia tensor with respect to the global coordinate frame can be computed as

$$\mathbf{I}(t) = \mathbf{R}(t)\mathbf{I}_0\mathbf{R}^T(t). \tag{6.21}$$

Matrix $\mathbf{I}(t)$ is in general time dependent, since object orientation relative to the global coordinate frame changes during the object's motion.

6.4.3 Linear and Angular Momentum

Linear momentum of a dimensionless particle with mass m is defined as

$$\mathbf{G}(t) = m\mathbf{v}(t). \tag{6.22}$$

If $\mathbf{v}(t)$ specifies the velocity of a rigid object center of mass, a similar equation can be written also for a rigid body

$$\mathbf{G}(t) = M\mathbf{v}(t). \tag{6.23}$$

A rigid body in this regard behaves similar to the mass particle with mass M.

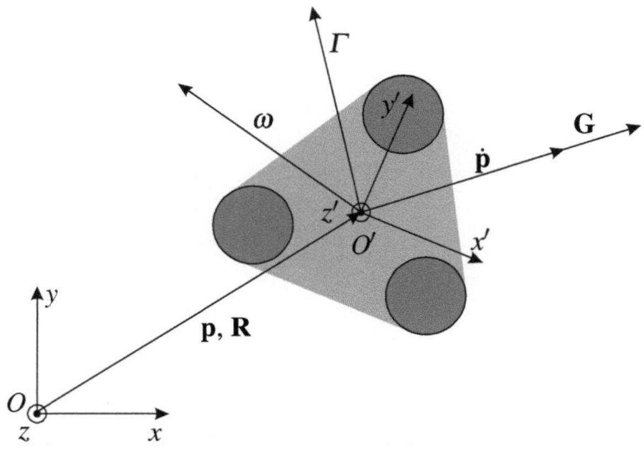

Fig. 6.11 Linear and angular momentum of a rigid object. The linear momentum vector is oriented in the direction of translational velocity. On the other hand, this is in general not the case for object angular momentum (the two vectors are aligned only if the object rotates about one of its principal axes)

If there is no external force acting on the body, the linear momentum is conserved and from Eq. (6.23) it is evident that also translational velocity of the object center of mass is constant.

A somewhat less intuitive concept than linear momentum is the object's angular momentum (angular momentum of a dimensionless object equals zero, since its moment of inertia equals zero). Angular momentum is defined by the product

$$\boldsymbol{\Gamma}(t) = \mathbf{I}(t)\boldsymbol{\omega}(t). \tag{6.24}$$

An object's angular momentum is conserved, if there is no external torque acting on the object. Conservation of angular momentum is an important reason for introducing this concept into the description of object dynamics. Namely, object rotational velocity $\boldsymbol{\omega}(t)$ may change even though the angular momentum is constant. Therefore, introduction of object angular momentum simplifies computation of equations of motion compared to the use of rotational velocities.

Vectors of linear and angular momentum are shown in Fig. 6.11.

6.4.4 Forces and Torques Acting on a Rigid Body

Motion of an object in space depends on forces and torques acting on it. Relations are shown in Fig. 6.12. A force $\mathbf{F}$ acts on the object with lever $\mathbf{r}$ relative to the object's center of mass. Force $\mathbf{F}$ results in a change of object linear momentum. At the same time force $\mathbf{F}$ produces torque $\mathbf{r} \times \mathbf{F}$, that in addition to other possible

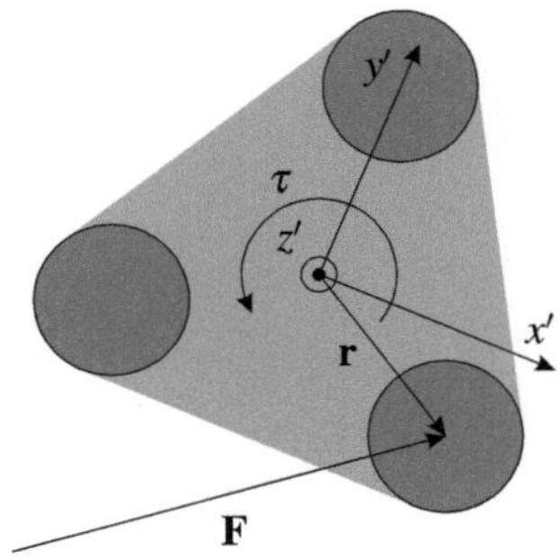

Fig. 6.12 Forces and torques acting on a rigid body

external torques, contributes to the overall torque $\boldsymbol{\tau}$ acting on the body and causes changes of the object's angular momentum.

The change of linear momentum equals the impulse of the sum of all forces $\mathbf{F}$ acting on the body, thus

$$d\mathbf{G}(t) = \mathbf{F}(t)dt. \tag{6.25}$$

The above equation can be rewritten as

$$\frac{d\mathbf{G}(t)}{dt} = \dot{\mathbf{G}}(t) = M\dot{\mathbf{v}}(t) = \mathbf{F}(t), \tag{6.26}$$

meaning that the time derivative of linear momentum equals the product of object mass and acceleration or the sum of all forces acting on the body.

The change of angular momentum equals the impulse of the sum of all torques $\boldsymbol{\tau}$ acting on the body, thus

$$d\boldsymbol{\Gamma}(t) = \boldsymbol{\tau}(t)dt. \tag{6.27}$$

The time derivative of angular momentum, therefore equals the sum of all torques acting on the body

$$\frac{d\boldsymbol{\Gamma}(t)}{dt} = \dot{\boldsymbol{\Gamma}}(t) = \boldsymbol{\tau}(t). \tag{6.28}$$

If forces and torques acting on the body are known, also time derivatives of linear and angular momenta are defined. Object linear momentum can thus be computed as a time integral of all forces acting on the body

$$\mathbf{G}(t) = \mathbf{G}_0 + \int_0^t \mathbf{F}(\xi)d\xi, \tag{6.29}$$

where $\mathbf{G}_0$ is the initial linear momentum.

Angular momentum can be computed as a time integral of all torques acting on the body

$$\boldsymbol{\Gamma}(t) = \boldsymbol{\Gamma}_0 + \int_0^t \boldsymbol{\tau}(\xi)d\xi, \tag{6.30}$$

where $\boldsymbol{\Gamma}_0$ indicates the initial angular momentum.

From the known linear $\mathbf{G}(t)$ and angular $\boldsymbol{\Gamma}(t)$ momenta and by taking into account object mass and inertia properties it is possible to compute object translational velocity $\mathbf{v}(t)$ and rotational velocity $\boldsymbol{\omega}(t)$ based on Eqs. (6.23) and (6.24), respectively.

In the following paragraphs we will introduce some typical forces and torques acting on an object in a virtual environment. These can be in general divided into five relevant contributions: (1) forces and torques as a result of interactions with the user (via haptic interface), (2) force field (gravity, magnetic field), (3) forces and torques produced by the medium through which the object moves (viscous damping), (4) forces resulting from interactions of an object with other objects in a virtual environment (collisions) and (5) virtual actuators (sources of forces and torques).

Interaction between a user and a virtual environment is often enabled by the use of virtual tools that the user manipulates through a haptic interface. Virtual actuators are sources of constant or variable forces and torques acting on an object. Virtual actuators can be models of electrical, hydraulic, pneumatic actuator systems, or, for example, engines with internal combustion. In this group we can also include biological actuators—muscles. Magnitude of forces and torques of virtual actuators can change automatically based on the unfolding of events within a virtual environment, or through interactions with the user.

The force field within which the objects move can be homogeneous (local gravity field) or nonhomogeneous (magnetic dipole filed). The force acting on the object depends on the field parameters and object properties. For example, a gravity force can be computed as (Fig. 6.13)

$$\mathbf{F}_g = M\mathbf{g}, \tag{6.31}$$

where $\mathbf{g}$ is the gravity acceleration vector. In this case the force is independent of the object motion. The homogeneous force field also does not cause any torque that would result in a change of object angular momentum.

Analysis of interaction forces between the object and the medium through which the object moves (Fig. 6.14) is relatively straightforward. In this case friction forces are of primary interest. In the case of a simple model based on viscous damping (an object floating in a viscous fluid), the interaction force can be computed as

$$\mathbf{F}_B = -B\mathbf{v}, \tag{6.32}$$

where B is the coefficient of viscous damping and $\mathbf{v}$ indicates object velocity. Friction forces oppose object motion, therefore a negative sign is introduced.

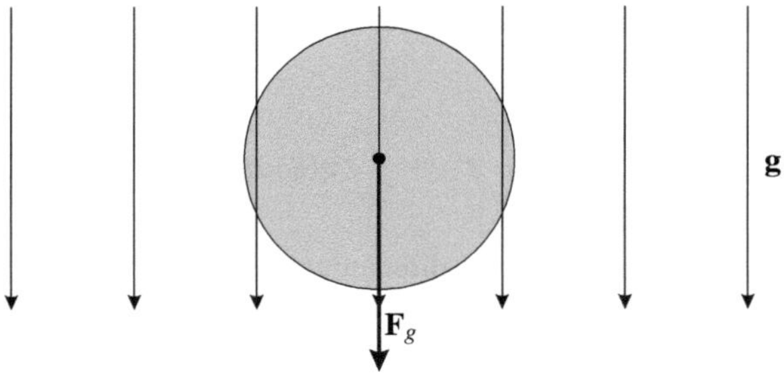

Fig. 6.13 Gravity field

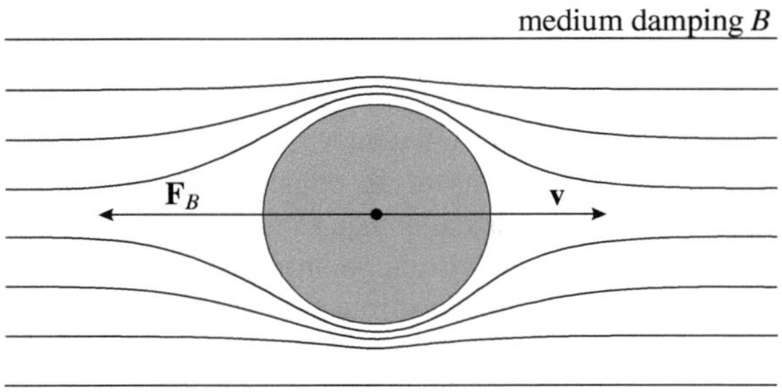

Fig. 6.14 Object in a viscous medium

The most complex analysis is that of forces and torques resulting from interactions between objects. A prerequisite is the implementation of an algorithm for collision detection (see Chap. 5). Then based on object dynamic properties (plasticity, elasticity) it is possible to compute reaction forces.

Relations during a collision are shown in Fig. 6.15. As a result of a collision between two objects, a plastic or elastic deformation occurs (the maximal deformation is indicated with d in the figure). Deformations depend on object stiffness properties. Object collisions and deformations need to be computed in real time. The computed deformations can be used for modeling interaction forces or for visualization purposes.

For the simplest case, when an object is modeled only as a spring with stiffness value k, the collision reaction force $\mathbf{F}_c$ can be computed as

$$\mathbf{F}_c = kd\mathbf{n}, \tag{6.33}$$

where d determines the object deformation and vector $\mathbf{n}$ determines the reaction force direction. For a simple case of a frictionless contact, the vector $\mathbf{n}$ can be determined as a normal vector to the surface of the object at a point of contact.

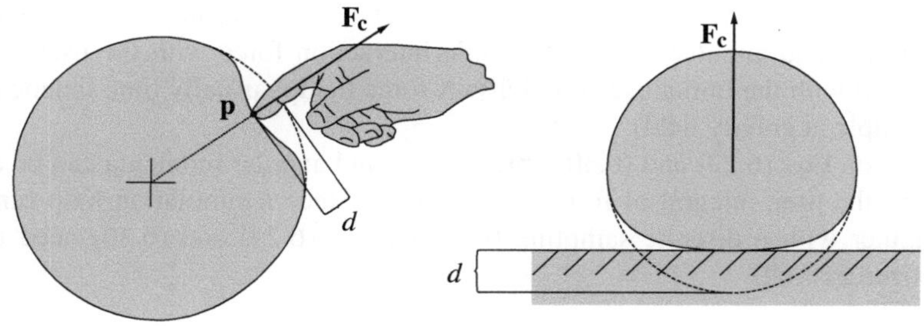

Fig. 6.15 Collision of two objects and a collision between an object and a grounded wall; **p**—contact point, d—object deformation (penetration) and $\mathbf{F}_c$—computed reaction force

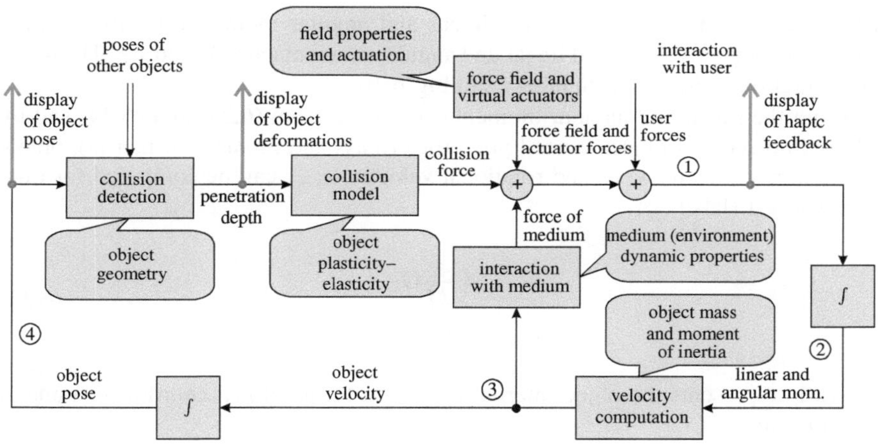

Fig. 6.16 Algorithm for computation of object motion as a result of interactions with other objects, with the user and with the medium. The force represents a generalized quantity that includes also torques

6.4.5 Computation of Object Motion

After an introduction of the basic equations of motion, we can now implement an algorithm for computation of object motion in a computer generated environment. A block scheme representing the basic concept is shown in Fig. 6.16. It is possible to note that the algorithm consists of a closed loop that needs to be computed within a single computational time step of the haptic rendering loop.

Since the loop in Fig. 6.16 is closed, there is no natural entry point. Therefore, we will follow labels in the order from ① to ④. We assume the force to be the cause of object motion. Point ① represents the sum of all forces and torques (user, virtual actuators, medium, collision, field) acting on the object. Forces resulting from possible collisions with other objects can be computed based on the initial pose of the object, determined with $\mathbf{p}_0$ and q_0, interaction force with the medium can be computed based on the initial object velocity (that can be determined from the initial object linear and angular momenta), while interaction force with the user can be determined with the initial user's activity. A force field is usually time independent (for example, a gravity field).

Based on Eqs. (6.29) and (6.30) object linear and angular momenta can be computed by the time integral of forces and torques. Since a simulation loop runs on a computer with a discrete sampling time Δt, Eqs. (6.29) and (6.30) need to be discretized as well:

$$\begin{aligned} \mathbf{G}_{k+1} &= \mathbf{G}_k + \mathbf{F}_k \Delta t \\ \boldsymbol{\Lambda}_{k+1} &= \boldsymbol{\Lambda}_k + \boldsymbol{\tau}_k \Delta t, \end{aligned} \tag{6.34}$$

where $\mathbf{G}_k$ and $\boldsymbol{\Lambda}_k$ represent object linear and angular momenta at discrete time interval k, respectively. Initial linear and angular momenta are $\mathbf{G}_0$ and $\boldsymbol{\Lambda}_0$. The result of integration is marked with label ② in Fig. 6.16.

From the linear and angular momenta computed from (6.23) and (6.24) and by taking into account object inertial properties (mass and moment of inertia), object translational velocity $\mathbf{v}_{k+1}$ and rotational velocity $\boldsymbol{\omega}_{k+1}$ can be computed for time instant $k+1$ (label ③)

$$\begin{aligned} \mathbf{v}_{k+1} &= \frac{1}{M}\mathbf{G}_{k+1} \\ \boldsymbol{\omega}_{k+1} &= \mathbf{I}_k^{-1}\boldsymbol{\Gamma}_{k+1}, \end{aligned} \tag{6.35}$$

where $\mathbf{I}_k$ represents the object inertia in relation to the global coordinate frame at time instant k.

Based on Eqs. (6.12) and (6.16) the new object pose can be computed. Equations (6.12) and (6.16) are numerically integrated

$$\begin{aligned} \mathbf{p}_{k+1} &= \mathbf{p}_k + \mathbf{v}_{k+1}\Delta t \\ q_{k+1} &= q_k + \frac{1}{2}\Delta t\, \tilde{\omega}_{k+1} \otimes q_k, \end{aligned} \tag{6.36}$$

where $\mathbf{p}_k$ and q_k are the object position and orientation at time interval k, while initial position and orientation are determined with $\mathbf{p}_0$ and q_0. The new object pose is now computed (label ④) as a consequence of interactions with other objects, with the user, with the medium and due to the effects of virtual actuators. The loop for computation of object pose continues in point ①, after all interaction forces are computed. The loop presented in Fig. 6.16 needs to be implemented for all dynamic objects in the virtual environment.

References

1. Okamura, A.M., Smaby, N., Cutkosky, M.R.: An overview of dexterous manipulation. In: Proceedings of the IEEE International Conference on Robotics and Automation, pp. 255–262 (2000)
2. Salisbury, J.K., Brock, D., Massie, T., Swarup, N., Zilles, C.: Haptic rendering: programming touch interaction with virtual objects. In: Symposium on Interactive 3D Graphics, pp. 123–130. Monterey, USA (1995)
3. Basdogan, C., Srinivasan, M.A.: Haptic rendering in virtual environments. In: Handbook of Virtual Environments: Design, Implementation, and Applications, pp. 117–134. Lawrence Erlbaum Associates, New Jersey (2001)
4. Lawrence, D., Pao, L.Y., Salada, M.A., Dougherty, A.M.: Quantitative experimental analysis of transparency and stability in haptic interfaces. In: Proceedings of the ASME Dynamic Systems and Control Division, pp. 441–449 (1996)
5. Jones, L.A., Hunter, I.W.: A perceptual analysis of stiffness. Exp. Brain Res. **79**(1), 150–156 (1990)
6. Olsson, H., Åström, K., Wit, C.C.D., Gäfvert, M., Lischinsky, P.: Friction models and friction compensation. Eur. J. Control **4**, 176–195 (1998)
7. McMillan, A.: A non-linear friction model for self-excited vibrations. J. Sound Vib. **205**, 323–335 (1997)
8. Ferretti, G., Magnani, G., Rocco, P.: Single and multistate integral friction models. IEEE Trans. Autom. Control **49**, 2292–2297 (2004)
9. Richard, C., Cutkosky, M.: Friction modeling and display in haptic applications involving user performance. IEEE Int. Conf. Robot. Autom. **49**, 605–611 (2002)
10. Dahl, P.: A solid friction model. Technical Report, The Aerospace Corporation, El Segundo, CA (1968)
11. Jeon, S., Metzger, J., Choi, S., Harders, M.: Extensions to haptic augmented reality: modulating friction and weight. In: IEEE World Haptics Conference (WHC), pp. 227–232 (2011)
12. Salcudean, S., Vlaar, T.: On the emulation of stiff walls and static friction with a magnetically levitated input/output device. ASME J. Dyn. Syst. Meas. Control **119**, 127–132 (1997)
13. Hayward, V., Armstrong, B.: A new computational model of friction applied to haptic rendering. In: Corke, P., Trevelyan, J. (eds.) Experimental Robotics VI, vol. 250, pp. 403–412. Springer-Verlag (2000)
14. Gillespie, R.B., Colgate, J.E.: A survey of multibody dynamics for virtual environments. In: Proceedings of the ASME IMECE, pp. 45–54. Dallas (1997)
15. Barraff, D.: An introduction to physically based modeling: rigid body simulation 1—unconstrained rigid body dynamics. SIGGRAPH Course Notes (1997)
16. Barraff, D.: An introduction to physically based modeling: rigid body simulation 2—nonpenetration constraints. SIGGRAPH Course Notes (1997)

Chapter 7
Control of Haptic Interfaces

In Chaps. 5 and 6 we analyzed collision detection and methods for modeling haptic contact occurring in virtual or real environments. Forces and movements that are either modeled or measured can be used as an input signal to the controller of the haptic interface. Selection of control strategy (impedance or admittance control) depends on the available hardware and software architectures as well as on the planned use of the haptic interface.

Interaction between the user and the environment presents a bilateral transfer of energy, as the product of force and displacement defines the mechanical work. The rate of change of energy (mechanical power) is defined by the instantaneous product of the interaction force and the movement velocity. The exchange of mechanical energy between the user and the haptic interface is the main difference compared to other display modalities (visual, acoustic) that are based on one-way flow of information with negligible energy levels.

If energy flow is not properly controlled, the effect of haptic feedback can be degraded, due to unstable behavior of the haptic device. Important issues related to control of haptic interaction are its quality and especially stability of haptic interaction, while taking into account properties of the human operator, who is inserted into the control loop [1, 2].

Two classes of control schemes for control of haptic interfaces can be defined: (1) impedance control providing force feedback and (2) admittance control providing displacement feedback.

Impedance approach to display of kinesthetic information is based on measuring user's motion velocity or limb position and implementation of a force vector at the point of measurement of position or velocity. We will assume that the point of interaction is the user's arm. Even though it is possible to construct kinesthetic displays also for other parts of the body, the arm is the primary human mechanism for precise manipulation tasks. The magnitude of the displayed force is determined through interaction with the virtual or remote (teleoperation) environment. In case of a virtual environment the force is computed as a response of a simulated object to displacement measured on the user's side of the haptic interface. In case of teleoperation,

M. Mihelj and J. Podobnik, *Haptics for Virtual Reality and Teleoperation*, Intelligent Systems, Control and Automation: Science and Engineering 64,
DOI: 10.1007/978-94-007-5718-9_7,

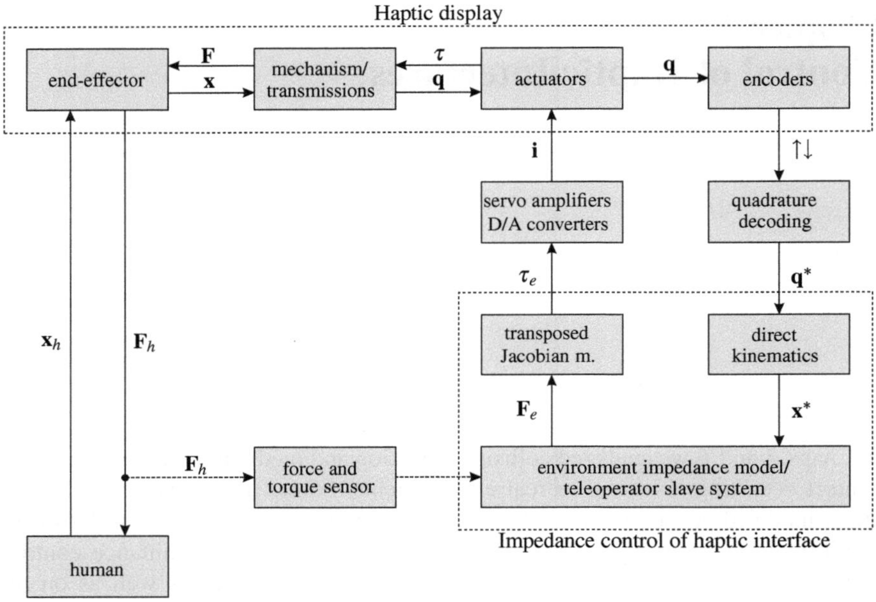

Fig. 7.1 Block scheme of impedance controlled haptic interface. *Arrows* indicate dominant direction of flow of information. *Hatched line* indicates supplementary information. Adapted from [3]

the object model is replaced by a robot slave system that tracks measured user's motion and the interaction force between the robot and the environment is used for controlling the haptic interface.

Figure 7.1 shows a block scheme of impedance controlled haptic interface. Joint position encoders measure angular displacements $\mathbf{q}^*$. These are then used in the forward kinematic model for computing the pose of the haptic interface end-effector $\mathbf{x}^*$. Based on the physical model of the environment and the haptic interface end-effector pose (interaction force between the user and the interface can be used as an additional input), desired reaction forces $\mathbf{F}_e$ are computed. Desired force is then transformed into the desired joint torques through the manipulator-transposed Jacobian matrix, and haptic display actuators generate desired torques. Actuator torques result in a haptic display end-effector force that is perceived by the user. A similar block scheme is valid also for the teleoperation system, where the environment model is replaced with a teleoperator slave device. Thus, the haptic interface displays forces that are being generated on the slave device of the teleoperation system or interaction forces being produced in the virtual environment.

The main characteristics of impedance display can be summarized as [4]:

- it has to enable unobstructed movement of the user arm, when there is no contact with the environment,
- it has to exactly reproduce forces that need to be applied on the user,

- it has to generate large forces in order to simulate or reproduce contacts with stiff objects and
- it has to have a wide bandwidth in order to reproduce transient responses during collisions with objects with sufficient fidelity and accuracy.

The above listed requirements are similar to the requirements for robots that are in contact with the environment and where interaction forces between the robot and the environment need to be precisely controlled.

An opposite approach to controlling haptic interfaces is based on measurement of forces, which the user applies on the haptic interface and display of displacements through the programmed motion of the haptic interface. The applied displacement is defined based on interactions within the virtual or remote environments. In a virtual environment the displacement is computed based on a dynamic model of the environment as a response of the environment to the measured force applied by the user. In a remote environment the remote robot manipulator is programmed to apply forces on the environment proportional to the force applied by the human operator on the haptic interface end-effector. The measured displacement of the slave system is used for controlling the displacement of the haptic interface.

Figure 7.2 shows the block scheme of the admittance controlled haptic interface. Joint encoders measure displacements $\mathbf{q}^*$. These are then used in the forward kinematic model of the haptic display for computation of end-effector pose $\mathbf{x}^*$. Based on

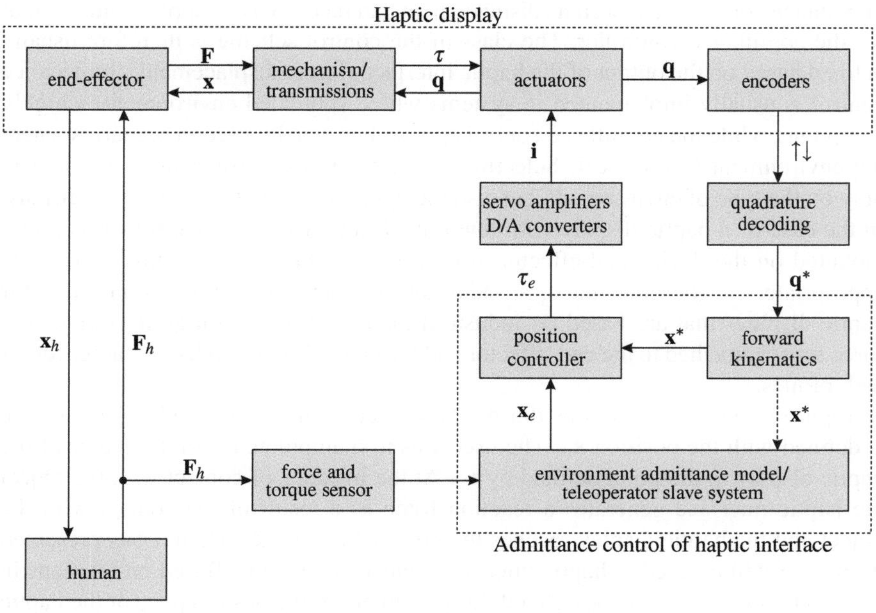

Fig. 7.2 Block scheme of admittance controlled haptic interface. *Arrows* indicate dominant direction of flow of information. *Hatched line* indicates supplementary information

the physical model of the environment and the force being applied by the user on the robot end-effector (information about the haptic interface pose can be used as supplementary information), desired haptic interface end-effector displacements $\mathbf{x}_e$ can be determined. These are then used as a reference input for the device position controller, of which output are the desired joint torques that must be generated by the haptic interface actuators. These generate movement of the haptic display that is finally perceived by the human operator. Similar block scheme can be designed also for the teleoperation system, where the environment model is replaced with the slave device of the teleoperation system. Thus, a haptic interface displays either displacements of the slave system in a teleoperation environment or displacements resulting from interactions with the virtual environment.

The main characteristics of admittance display can be summarized as [4]:

- mechanism needs to be stiff enough to completely prevent movement of the user's arm when in contact with a stiff object,
- it has to exactly reproduce desired displacement,
- it has to be backdrivable to allow reproduction of free movement and
- bandwidth of the system needs to be large enough to allow reproduction of transient responses with sufficient fidelity and accuracy.

The above mentioned characteristics are similar to the characteristics of position controlled robot manipulators, where high accuracy of positional tracking is required.

In some cases the interaction force can be used as an additional input to the impedance controller, as well as displacement can be used as a supplementary input for the admittance controller. The class of the control scheme is therefore usually defined based on the output of the haptic interface (force, displacement). Impedance control is usually implemented in systems where simulated environment is highly compliant, while the admittance control approach is usually used in scenarios where the environment is very stiff. Selection of the type of controller does not depend only on the type of environment, but also the dynamic properties of a haptic display. In the case of a haptic display with low impedance, where a force sensor is rarely mounted on the device end-effector, use of an impedance type controller is more appropriate. On the other hand, an admittance controller is more appropriate for haptic displays that are based on industrial manipulators with high impedances, a force sensor attached to the end-effector and the capability of producing larger forces and torques.

Figure 7.3 shows conditions during the contact. An undeformed object surface is defined with the position $\mathbf{x}_0$. The user tries to manipulate the end-effector of the haptic display to the point defined by $\mathbf{x}_e$. At the instance of contact with the object the haptic interface generates a reaction force as a result of the contact with the object and the haptic display reaches the pose defined by $\mathbf{x}_h$. Definitions presented in Fig. 7.3 will be used in haptic interface control algorithms. Based on relations in Fig. 7.3 two variables can be defined. Displacement of the desired pose of the haptic display from the pose of the undeformed object surface can be defined as

$$\Delta\mathbf{x}_e = \mathbf{x}_e - \mathbf{x}_0 \tag{7.1}$$

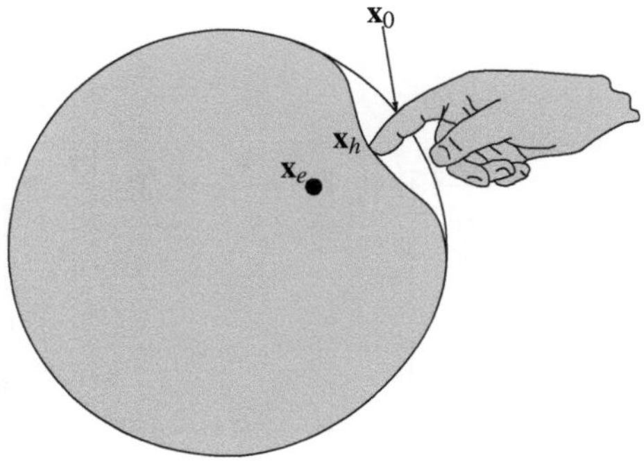

Fig. 7.3 Interaction with a compliant environment

and displacement of the actual pose of the haptic display relative to the undeformed object surface is defined as

$$\Delta \mathbf{x}_h = \mathbf{x}_h - \mathbf{x}_0. \tag{7.2}$$

In addition to the two displacements defined above, two velocity variables will be introduced. Variable $\mathbf{v}_e = \Delta \dot{\mathbf{x}}_e$ defines the desired velocity of the object surface deformation and $\mathbf{v}_h = \Delta \dot{\mathbf{x}}_h$ defines the actual velocity of the object surface deformation.

7.1 Open-Loop Impedance Control

The case where there is no force feedback loop from the haptic display to the device controller is referred to as an open-loop impedance control of a haptic interface [2, 3]. Accuracy of force control in this case mainly depends on the dynamics of the haptic display that has to be negligible compared to the dynamics of the simulated environment. Friction, gravity and inertial forces originating from the device are directly added to the forces resulting from haptic rendering and are perceived by the user. Haptic display geometric model errors and errors in haptic actuators are also transformed into a force acting on the haptic display end-effector. The advantage of using an open-loop impedance controller mainly lies in its simplicity and the fact that there is no need for an external force and torque sensor, which significantly reduces complexity, cost and weight of the device.

An example of an open-loop impedance controller for a haptic interface is shown in Fig. 7.4. Linearized dynamics of a haptic display is given by the impedance of a haptic display in joint space ${}^j\mathbf{Z}_r$, which is defined as the ratio between joint torque $\boldsymbol{\tau}$ and joint velocity $\dot{\mathbf{q}}$, meaning $\boldsymbol{\tau} = {}^j\mathbf{Z}_r\dot{\mathbf{q}}$. In order to simplify the analysis of a haptic interface controller, the movement of the haptic display will be defined in velocity terms instead of positions. In this case the relation between joint velocities and end-effector

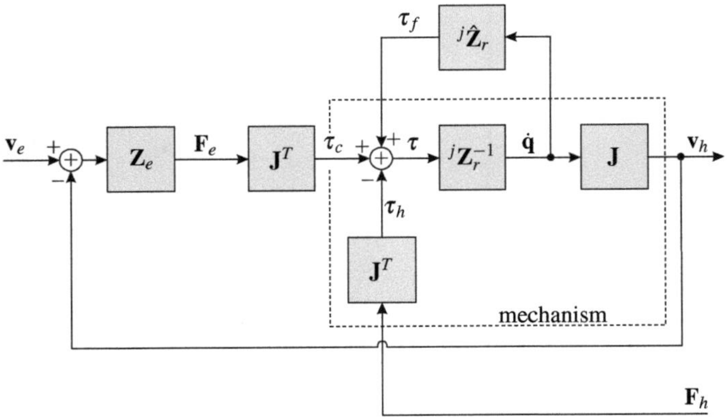

Fig. 7.4 Open-loop impedance control of a haptic interface

velocities is defined by the Jacobian matrix $\mathbf{J}$ of the haptic display. Based on the relations in Fig. 7.4 the input to the controller is defined as the difference between the desired end-effector velocity $\mathbf{v}_e$ and actual end-effector velocity $\mathbf{v}_h$. The desired force at the haptic display end-effector $\mathbf{F}_e$ is computed from the difference of velocities and impedance of the virtual environment $\mathbf{Z}_e$. The computed force is then transformed to the desired joint torques $\boldsymbol{\tau}_c$ via multiplication with the manipulator transposed Jacobian matrix. $\mathbf{F}_h$ is the force that the user applies on the haptic display end-effector and therefore represents a physical and not a control input to the haptic interface. Variable $\boldsymbol{\tau}_f$ defines the feedforward torque that actively compensates haptic display intrinsic dynamics. The sum of all joint torques is transformed to joint velocities $\dot{\mathbf{q}}$ via the matrix of haptic display joint impedances ${}^j\mathbf{Z}_r$, finally resulting in haptic display end-effector velocity $\mathbf{v}_h$.

For the impedance control approach shown in Fig. 7.4, the ratio between the force $\mathbf{F}_h$ and end-effector velocity $\mathbf{v}_h$ will be determined. The ratio defines the system impedance perceived by the user. The following two relations can be written directly from the block scheme

$$\mathbf{v}_h = \mathbf{J}{}^j\mathbf{Z}_r^{-1}(\mathbf{J}^T\mathbf{F}_e - \mathbf{J}^T\mathbf{F}_h + {}^j\hat{\mathbf{Z}}_r\dot{\mathbf{q}}), \tag{7.3}$$

$$\mathbf{F}_e = \mathbf{Z}_e(\mathbf{v}_e - \mathbf{v}_h). \tag{7.4}$$

By inserting Eq. (7.4) into (7.3) and replacing $\dot{\mathbf{q}}$ by $\mathbf{J}^{-1}\mathbf{v}_h$, we obtain

$$\mathbf{v}_h = \mathbf{J}{}^j\mathbf{Z}_r^{-1}\left[\mathbf{J}^T\mathbf{Z}_e(\mathbf{v}_e - \mathbf{v}_h) - \mathbf{J}^T\mathbf{F}_h + {}^j\hat{\mathbf{Z}}_r\mathbf{J}^{-1}\mathbf{v}_h\right]. \tag{7.5}$$

We premultiply expression ${}^j\hat{\mathbf{Z}}_r\mathbf{J}^{-1}\mathbf{v}_h$ with $\mathbf{J}^T\mathbf{J}^{-T}$ and expose $\mathbf{J}^T$ from brackets. We define the intrinsic dynamics of the haptic interface expressed in the task space $\mathbf{Z}_r$ as an equivalent of the intrinsic dynamics expressed in the joint space

($\boldsymbol{\tau} = {}^j\mathbf{Z}_r\dot{\mathbf{q}} \quad \leftrightarrow \quad \mathbf{F} = \mathbf{Z}_r\mathbf{v}$). Impedance $\mathbf{Z}_r$ can be computed as

$$\begin{array}{ccccc} \boldsymbol{\tau} = \mathbf{J}^T\mathbf{F} \quad \mathbf{v} = \mathbf{J}\dot{\mathbf{q}} & & \boldsymbol{\tau} = {}^j\mathbf{Z}_r\dot{\mathbf{q}} & & \\ \downarrow \qquad \downarrow & & \downarrow & & \text{substitutions} \\ \mathbf{F} = \mathbf{Z}_r\mathbf{v} & \Rightarrow & \mathbf{J}^{-T}\boldsymbol{\tau} = \mathbf{Z}_r\mathbf{J}\dot{\mathbf{q}} & \Rightarrow & \mathbf{J}^{-T}\,{}^j\mathbf{Z}_r\dot{\mathbf{q}} = \mathbf{Z}_r\mathbf{J}\dot{\mathbf{q}} \\ & & & & \Downarrow \\ & & & & \mathbf{Z}_r = \mathbf{J}^{-T}\,{}^j\mathbf{Z}_r\mathbf{J}^{-1}. \end{array} \tag{7.6}$$

The ratio between the force $\mathbf{F}_h$ and velocity $\mathbf{v}_h$ will be computed with the assumption that the desired velocity equals zero, $\mathbf{v}_e = 0$. Equation (7.5) can be rewritten as

$$\mathbf{v}_h = \mathbf{Z}_r^{-1}[-\mathbf{Z}_e\mathbf{v}_h - \mathbf{F}_h + \hat{\mathbf{Z}}_r\mathbf{v}_h]. \tag{7.7}$$

With a simple algebraic operation it is possible to compute the ratio between the force acting on the user $-\mathbf{F}_h$ and the velocity $\mathbf{v}_h$, which results in the impedance of the haptic interface perceived by the user

$$\mathbf{Z}_{cl} = \frac{-\mathbf{F}_h}{\mathbf{v}_h} = \mathbf{Z}_e + \mathbf{Z}_r - \hat{\mathbf{Z}}_r. \tag{7.8}$$

It is clear from Eq. (7.8) that, when the intrinsic dynamics of the haptic display is not compensated, the impedance perceived by the user equals the sum of impedances of the virtual environment $\mathbf{Z}_e$ and intrinsic impedance of the haptic display expressed in the task space

$$\mathbf{Z}_{cl} = \mathbf{Z}_e + \mathbf{Z}_r. \tag{7.9}$$

Since the control goal is to make the user perceive only the virtual environment impedance, it is necessary to reduce the effect of the intrinsic dynamics by adapting the haptic display mechanism. This usually means that it is necessary to reduce friction and mass properties of the mechanism. The controller can also be improved by introducing the compensation of the device intrinsic impedance $\hat{\mathbf{Z}}_r$, meaning that it is possible to reduce the effect of the device intrinsic dynamics. The feedback compensation of the device intrinsic dynamics can improve the performance of the device, however it can also cause problems. The controller becomes susceptible to model errors; if estimated friction, mass and inertia parameters are not correct, $\hat{\mathbf{Z}}_r \neq \mathbf{Z}_r$, the output impedance will not be correct as well.

7.2 Closed-Loop Impedance Control

A block scheme for the closed-loop impedance control is shown in Fig. 7.5. By comparing the new control approach with the block scheme Fig. 7.4, it is possible to note a difference in the loop that leads from the force sensor to the impedance controller. The force feedback loop closes the loop for controlling the desired force through the proportional force gain matrix $\mathbf{K}_F$ [3, 5, 6]. As for the open-loop control approach, we will again compute the output impedance, the impedance perceived by the user of the haptic interface. Based on the relations in Fig. 7.5 we can write the following equations

$$\mathbf{v}_h = \mathbf{J}^j\mathbf{Z}_r^{-1}(\boldsymbol{\tau}_c - \mathbf{J}^T\mathbf{F}_h + {}^j\hat{\mathbf{Z}}_r\dot{\mathbf{q}}), \tag{7.10}$$

$$\boldsymbol{\tau}_c = \mathbf{J}^T\,(\mathbf{F}_e + \mathbf{K}_F(\mathbf{F}_e - \mathbf{F}_h)), \tag{7.11}$$

$$\mathbf{F}_e = \mathbf{Z}_e(\mathbf{v}_e - \mathbf{v}_h). \tag{7.12}$$

The ratio between the force $\mathbf{F}_h$ and velocity $\mathbf{v}_h$ will again be computed based on the assumption that $\mathbf{v}_e = 0$. Equation (7.12) will be inserted into Eq. (7.11) and the result will be inserted into (7.10). By considering the definition of the impedance of the haptic display expressed in the task space (7.6) we obtain

$$\mathbf{v}_h = \mathbf{Z}_r^{-1}\left[-(\mathbf{I} + \mathbf{K}_F)\mathbf{Z}_e\mathbf{v}_h - (\mathbf{I} + \mathbf{K}_F)\mathbf{F}_h + \hat{\mathbf{Z}}_r\mathbf{v}_h\right], \tag{7.13}$$

where $\mathbf{I}$ is an identity matrix with dimensions equal to the dimensions of the task space. With a simple algebraic operation it is possible to compute the ratio between the force applied by the user on the haptic display end-effector $-\mathbf{F}_h$ and the velocity

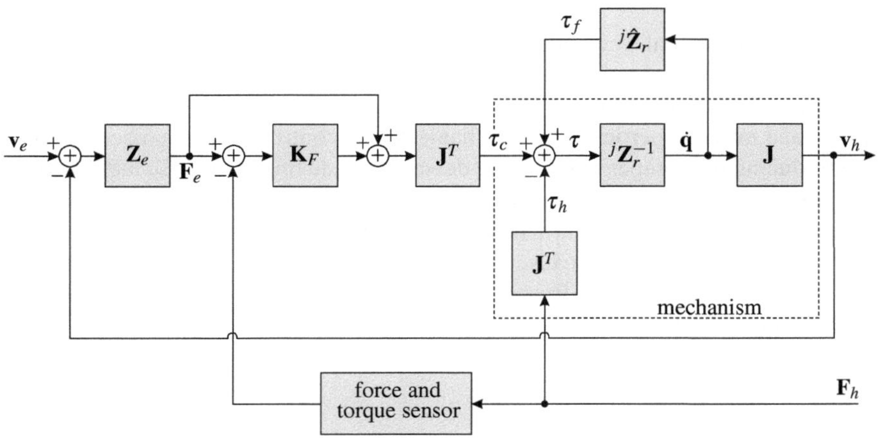

Fig. 7.5 Closed-loop impedance control of a haptic interface

$\mathbf{v}_h$. The ratio defines the output impedance of the haptic interface being perceived by the operator

$$\mathbf{Z}_{cl} = \frac{-\mathbf{F}_h}{\mathbf{v}_h} = \mathbf{Z}_e + (\mathbf{I} + \mathbf{K}_F)^{-1}(\mathbf{Z}_r - \hat{\mathbf{Z}}_r). \tag{7.14}$$

It is possible to note from Eq. (7.14) that the error of the output impedance is proportional to the inverse value of the force gain matrix $\mathbf{K}_F$. If the force gain matrix equals zero, the control approach becomes equivalent to the open-loop impedance control. Force control gains should be defined as high as possible, where their upper limit is constrained by the stability conditions of the haptic interface.

Next, we will analyze the effect of human arm dynamics on the closed-loop impedance control. We will consider the dynamic model of the human arm as presented in Sect. 3.4. A block scheme that includes elements of haptic interaction shown in Figs. 7.5 and 3.6a is presented in Fig. 7.6. The closed-loop impedance control scheme has been simplified by taking into account the relations $\mathbf{v}_e = 0$, $\mathbf{Z}_r = \mathbf{J}^{-T\,j}\mathbf{Z}_r\mathbf{J}^{-1}$ (Eq. 7.6) and $\hat{\mathbf{Z}}_r = \mathbf{J}^{-T\,j}\hat{\mathbf{Z}}_r\mathbf{J}^{-1}$.

We are interested in the ratio between the central nervous system commands $\mathbf{u}$ and the movement velocity of the contact point between the user and the haptic display end-effector. For this purpose it is possible to simplify the block scheme in Fig. 7.6 into the simple block scheme shown in Fig. 7.7. The simplification can be done by taking into account the simplified model of human arm dynamics shown in Fig. 3.6b and Eq. (7.14) that defines the closed-loop impedance of the haptic interface. The closed-loop impedance defines the ratio between the force that the user applies at the haptic display end-effector and the velocity of the end-effector motion. As already

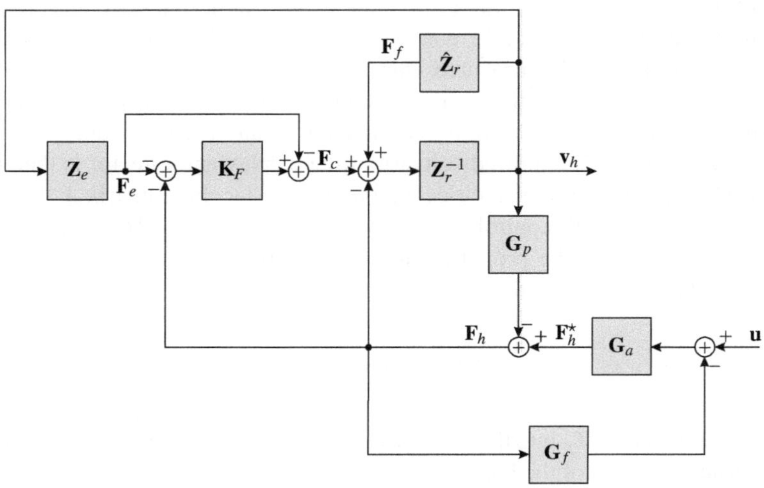

Fig. 7.6 Closed-loop impedance control of a haptic interface extended with the dynamic model of the operator's arm

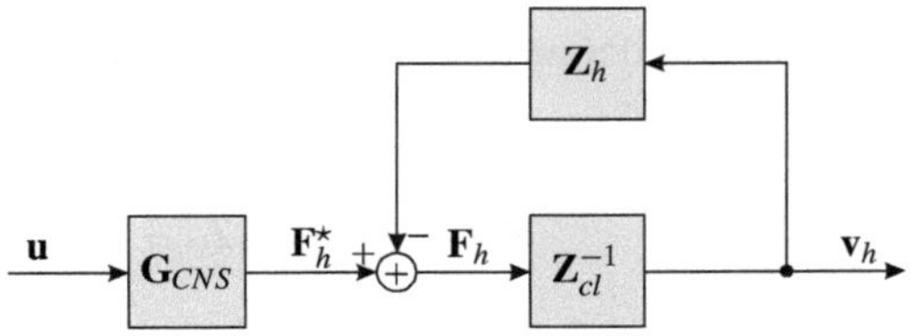

Fig. 7.7 A simplified model of the closed-loop impedance control of a haptic interface with the operator's arm dynamics taken into account

discussed in Sect. 3.4, the force acting on the end-effector depends on the central nervous system commands $\mathbf{u}$, conveyed across transfer function $\mathbf{G}_{CNS}$ (Eq. 3.17) and the velocity of the movement of the haptic display, which, mapped through the biomechanical impedance of the human arm $\mathbf{Z}_h$ (Eq. 3.18), reduces the force with which the user acts on the haptic display end-effector.

The block scheme shown in Fig. 7.7 indicates that when modeling haptic interaction, it is necessary to take into account mechanical parameters of the operator's arm, characteristics of the central nervous system and closed-loop impedance of the haptic interface.

7.3 Closed-Loop Admittance Control

Admittance controllers (with the measured interaction force as the input) were first developed for robotic applications. However, the approach that satisfies the requirements of haptic interactions is conceived as position-based impedance control [3, 5]. As shown in Fig. 7.8, the controller is based on two feedback loops: the external feedback loop for controlling the impedance of the haptic interface and the internal joint based feedback loop for controlling haptic display joint positions. In the external feedback loop the measured interaction force between the display and the user is used for computing the desired velocity of the haptic display end-effector by taking into account the controller admittance $\mathbf{Y}_c$ and the impedance of the virtual environment $\mathbf{Z}_e$. The desired velocity of the display end-effector is then transformed into the joint space velocities via the inverse Jacobian matrix $\mathbf{J}^{-1}$. A local proportional-integral (PI) controller in each haptic display joint ${}^j\mathbf{D}$ (PI velocity controller is equivalent to the proportional-derivative position controller) guarantees tracking of the desired joint trajectories

$$ {}^j\mathbf{D} = \mathbf{K}_D + \frac{1}{s}\mathbf{K}_P. \tag{7.15}$$

With the inclusion of ${}^j\hat{\mathbf{Z}}_r$ into the feedback loop it is possible to partially compensate the intrinsic dynamics of the haptic display.

Next, we define the output impedance, the impedance perceived by the user of the haptic interface. From relations in Fig. 7.8 we can write the following relations

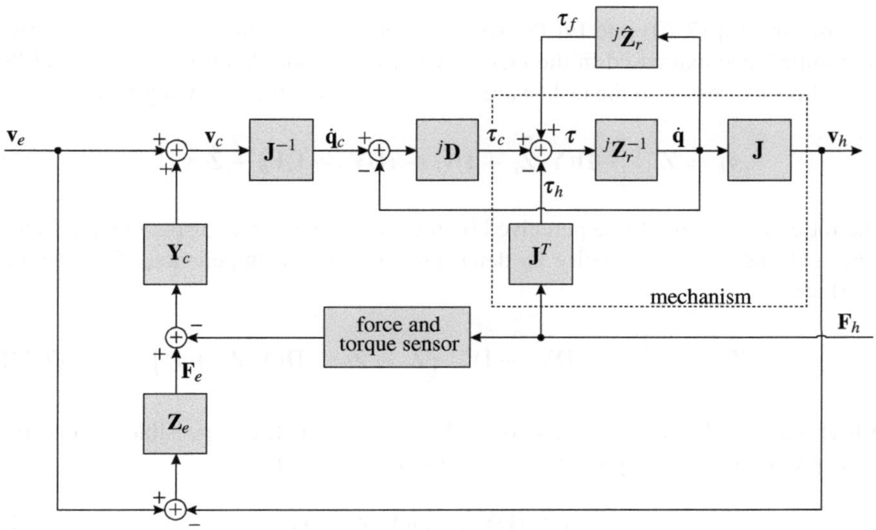

Fig. 7.8 Closed-loop admittance control of a haptic interface

$$\mathbf{v}_h = \mathbf{J}\,{}^j\mathbf{Z}_r^{-1}(\boldsymbol{\tau}_c - \mathbf{J}^T\mathbf{F}_h + {}^j\hat{\mathbf{Z}}_r\dot{\mathbf{q}}), \tag{7.16}$$

$$\boldsymbol{\tau}_c = {}^j\mathbf{D}(\dot{\mathbf{q}}_c - \dot{\mathbf{q}}), \tag{7.17}$$

$$\dot{\mathbf{q}}_c = \mathbf{J}^{-1}\left(\mathbf{v}_e + \mathbf{Y}_c(\mathbf{Z}_e(\mathbf{v}_e - \mathbf{v}_h) - \mathbf{F}_h)\right), \tag{7.18}$$

$$\dot{\mathbf{q}} = \mathbf{J}^{-1}\mathbf{v}_h. \tag{7.19}$$

The ratio between the force $\mathbf{F}_h$ and velocity $\mathbf{v}_h$ will be computed at $\mathbf{v}_e = 0$. By inserting Eqs. (7.18) and (7.19) into (7.17) we obtain

$$\boldsymbol{\tau}_c = {}^j\mathbf{D}\mathbf{J}^{-1}\left(-(\mathbf{Y}_c\mathbf{Z}_e + \mathbf{I})\mathbf{v}_h - \mathbf{Y}_c\mathbf{F}_h\right). \tag{7.20}$$

Next, we define transfer functions $\mathbf{D}$ of local PI controllers expressed in the task space, while taking into account duality $\boldsymbol{\tau}_c = {}^j\mathbf{D}\Delta\dot{\mathbf{q}} \quad \leftrightarrow \quad \mathbf{F}_c = \mathbf{D}\Delta\mathbf{v}$, where $\Delta\dot{\mathbf{q}} = \dot{\mathbf{q}}_c - \dot{\mathbf{q}}$ and $\Delta\mathbf{v} = \mathbf{v}_c - \mathbf{v} = \mathbf{J}\Delta\dot{\mathbf{q}}$:

$$\begin{array}{ccccc} \boldsymbol{\tau}_c = \mathbf{J}^T\mathbf{F}_c & \Delta\mathbf{v} = \mathbf{J}\Delta\dot{\mathbf{q}} & \boldsymbol{\tau}_c = {}^j\mathbf{D}\Delta\dot{\mathbf{q}} & & \\ \downarrow & \downarrow & \downarrow & & \text{substitutions} \\ \mathbf{F}_c = \mathbf{D}\Delta\mathbf{v} & \Rightarrow & \mathbf{J}^{-T}\boldsymbol{\tau}_c = \mathbf{D}\mathbf{J}\Delta\dot{\mathbf{q}} & \Rightarrow & \mathbf{J}^{-T}\,{}^j\mathbf{D}\Delta\dot{\mathbf{q}} = \mathbf{D}\mathbf{J}\Delta\dot{\mathbf{q}} \\ & & & & \Downarrow \\ & & & & \mathbf{D} = \mathbf{J}^{-T}\,{}^j\mathbf{D}\mathbf{J}^{-1} \end{array} \tag{7.21}$$

By inserting Eq. (7.20) into Eq. (7.16), considering the definition of the haptic display impedance expressed in the task space (7.6) and the definition of the local PI controllers expressed in the task space (7.21), we obtain the following relation

$$\mathbf{v}_h = \mathbf{Z}_r^{-1}\left[-\mathbf{D}(\mathbf{Y}_c\mathbf{Z}_e + \mathbf{I})\mathbf{v}_h - (\mathbf{D}\mathbf{Y}_c + \mathbf{I})\mathbf{F}_h + \hat{\mathbf{Z}}_r\mathbf{v}_h\right]. \tag{7.22}$$

The ratio between the force perceived by the user at the haptic display end-effector $-\mathbf{F}_h$ and end-effector velocity $\mathbf{v}_h$ determines the output impedance of the haptic interface

$$\mathbf{Z}_{cl} = \frac{-\mathbf{F}_h}{\mathbf{v}_h} = (\mathbf{D}\mathbf{Y}_c + \mathbf{I})^{-1}\left(\mathbf{Z}_r - \hat{\mathbf{Z}}_r + \mathbf{D}(\mathbf{Y}_c\mathbf{Z}_e + \mathbf{I})\right). \tag{7.23}$$

If high gains of local PI position controllers are assumed, it is possible to simplify the expression for the haptic interface output impedance to

$$\mathbf{Z}_{cl} \approx (\mathbf{D}\mathbf{Y}_c)^{-1}\mathbf{D}(\mathbf{Y}_c\mathbf{Z}_e + \mathbf{I}) \tag{7.24}$$

or

$$\mathbf{Z}_{cl} \approx \mathbf{Z}_e + \mathbf{Y}_c^{-1}. \tag{7.25}$$

It is possible to note that the impedance perceived by the user is defined by the impedance of the virtual environment $\mathbf{Z}_e$ and the inverse value of the controller admittance $\mathbf{Y}_c$. The latter can be used as part of the definition of virtual environment. If, on the other hand, the virtual environment is completely defined with the expression $\mathbf{Z}_e$, the term $\mathbf{Y}_c$ defines the admittance of the controller that relates the desired end-effector forces and velocities. In this case the admittance should be as high as possible in order not to distort the impedance image of the virtual environment. The virtual environment can also be completely defined by the admittance function $\mathbf{Y}_c$, where $\mathbf{Z}_e = \mathbf{0}$. In this case the forces are applied by the user on the haptic display end-effector, directly transformed into the end-effector desired velocities through the admittance filters $\mathbf{Y}_c$. The virtual environment in this case behaves as impedance $\mathbf{Y}_c^{-1}$.

Finally, we analyze the effect of human arm dynamics on the closed-loop admittance control. The dynamic model of the human arm as presented in Sect. 3.4 will be considered. A block scheme that includes elements of haptic interaction shown in Figs. 7.8 and 3.6a is shown in Fig. 7.9. The closed-loop admittance control scheme is simplified by taking into account the relations $\mathbf{v}_e = 0, \mathbf{Z}_r = \mathbf{J}^{-T\,j}\mathbf{Z}_r\mathbf{J}^{-1}$ (Eq. 7.6), $\hat{\mathbf{Z}}_r = \mathbf{J}^{-T\,j}\hat{\mathbf{Z}}_r\mathbf{J}^{-1}$ and $\mathbf{D} = \mathbf{J}^{-T\,j}\mathbf{D}\mathbf{J}^{-1}$ (Eq. 7.21).

We are interested in the ratio between the central nervous system commands $\mathbf{u}$ and the movement velocity of the contact point between the user and the haptic display end-effector. For this purpose it is possible to simplify the block scheme 7.9 into the block scheme shown in Fig. 7.10. The simplification is done by taking into account the model of human arm dynamics shown in Fig. 3.6b and Eq. (7.23) that defines the closed-loop impedance of the haptic interface. The closed-loop impedance defines

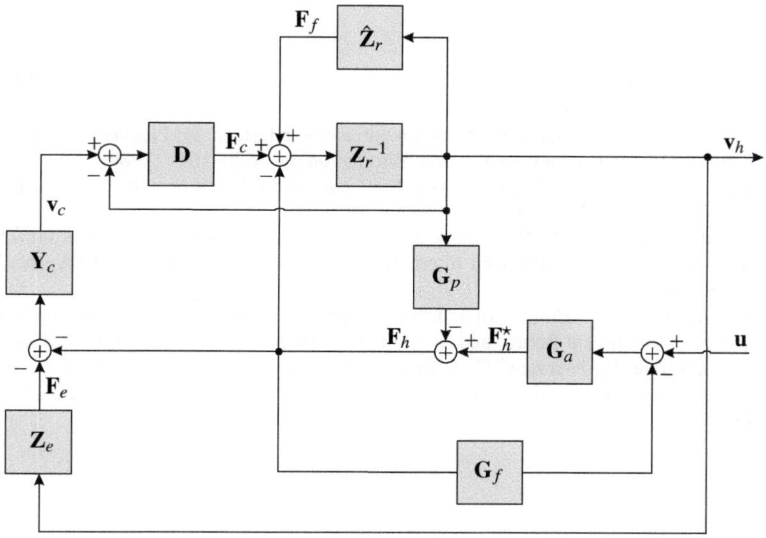

Fig. 7.9 Closed-loop admittance control of a haptic interface extended with the dynamic model of the operator's arm

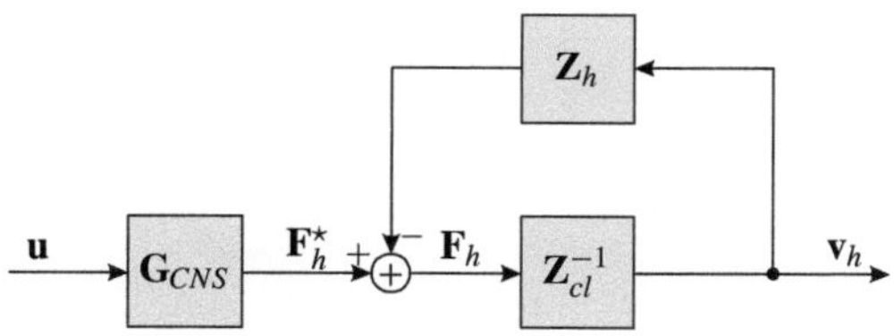

Fig. 7.10 A simplified model of the closed-loop admittance control of a haptic interface with the operator's arm dynamics taken into account

the ratio between the force that the user applies at the haptic display end-effector and the velocity of the end-effector motion. As already discussed in Sect. 3.4, the force acting on the end-effector depends on the central nervous system commands $\mathbf{u}$, conveyed across transfer function $\mathbf{G}_{CNS}$ (Eq. 3.17) and the velocity of the movement of the haptic display, which mapped through the biomechanical impedance of the human arm $\mathbf{Z}_h$ (Eq. 3.18) reduces the force with which the user acts on the haptic display end-effector.

The block scheme shown in Fig. 7.10 indicates that when modeling haptic interaction it is necessary to take into account mechanical parameters of the operator's arm, characteristics of the central nervous system and the closed-loop impedance of the haptic interface.

References

1. Kazerooni, H., Her, M.G.: The dynamics and control of a haptic interface device. IEEE Trans. Rob. Autom. **20**, 453–464 (1994)
2. Hogan, N.: Controlling impedance at the man/machine interface. In: Proceedings of the IEEE International Conference on Robotics and Automation pp. 1626–1631 (1989)
3. Carignan, C.R., Cleary, K.R.: Closed-loop force control for haptic simulation of virtual environments. Haptics-e **1**(2), 1–14 (2000)
4. Hannaford, B., Venema, S.: Kinesthetic Displays for Remote and Virtual Environments. In: Virtual environments and advanced interface design, pp. 415–436. Oxford University Press, New York (1995)
5. Ueberle, M., Buss, M.: Control of kinesthetic haptic interfaces. In: IEEE/RSJ International Conference on Intelligent Robots and Systems, Workshop on Touch and Haptics (2004)
6. Frisoli, A., Sotgiu, E., Avizzano, C., Checcacci, D., Bergamasco, M.: Force-based impedance control of a haptic master system for teleoperation. J. Dyn. Syst. Meas. Contr. **24**(1), 42–50 (2004)

Chapter 8
Stability Analysis of Haptic Interfaces

In the previous chapter we analyzed different methods for control of haptic interfaces. We addressed settings of controller gain parameters and the effect of user behavior on system performance. However, we did not address the stability criteria. Stability analysis of haptic interfaces is specific compared to the stability analysis of robotic devices, therefore the entire chapter is dedicated to this topic. We will analyze conditions leading to unstable behavior of a haptic interface and propose solutions for the stability problem.

8.1 Active Behavior of a Virtual Spring

When a haptic interface is used for rendering contact with a stiff virtual wall, annoying oscillatory behavior of a haptic display often occurs (unstable behavior). Such behavior distracts the user, since it is not natural for contacts in a real world [1]. Oscillations can be attributed to the supply of mechanical energy in a strongly coupled system (simulated virtual wall, haptic interface, human arm) through *energy leakage*.

Two aspects particular for haptic interactions significantly affect system stability. Both aspects result from implementation of a discretized virtual wall. It is known from control theory that the controller designed in a continuous space and implemented in a discrete space will not perform satisfactorily, if the controller sampling time is large compared to the dynamics of the controlled process. A nominal delay of half the sampling time related to the zero-order hold element results in a destabilizing effect. In the case of a stiff virtual wall, fast dynamic responses can be expected during the transitions from free space to the contact with the wall.

Figure 8.1 graphically illustrates the importance of the zero-order hold element in the control loop [2]. In order to simplify our analysis we will assume that the virtual wall is simply modeled as an ideal spring system, where the reaction force is defined by the product of spring stiffness and the depth of penetration into the wall. In the case

M. Mihelj and J. Podobnik, *Haptics for Virtual Reality and Teleoperation*, Intelligent Systems, Control and Automation: Science and Engineering 64, DOI: 10.1007/978-94-007-5718-9_8,

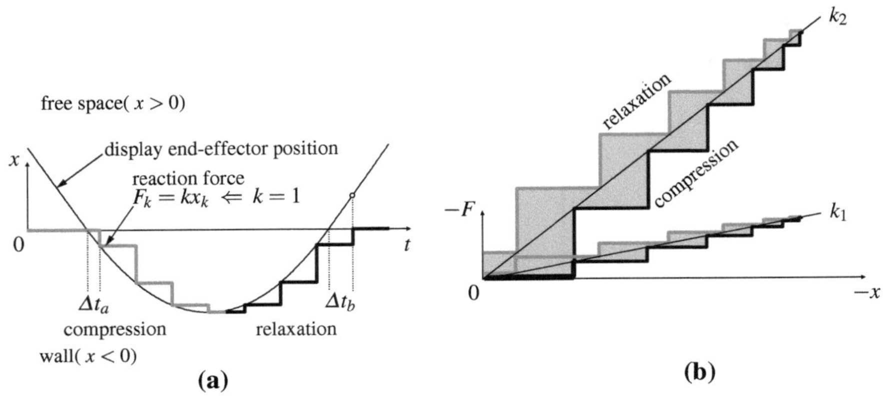

Fig. 8.1 Active behavior of a virtual spring

of a real world wall contact the curve that illustrates the ratio between the force and displacement is a continuous one. An ideal spring is an element without mechanical losses, therefore, all energy introduced into the system during the compression is then released from the system during the relaxation. However, due to the sampled nature of the discrete implementation of the spring, the conditions change. When moving into the wall, the sampled haptic display position will always be closer to the wall surface than the actual position of the display. Therefore, the reaction force will always be smaller than the reaction force of an ideal continuous spring that defines a real wall—Fig. 8.1a. On the other hand, when retracting from the virtual wall, the sampled haptic display position will always be deeper than the actual position of the haptic display. Therefore, the computed reaction force will be too large. By taking into account both conditions it becomes clear that during spring compression, less work is required for compression of a virtual spring compared to an ideal continuous spring of the same stiffness. When retracting from the wall, the returned mechanical work of a virtual spring will be larger than the work returned by the continuous spring. Thus, a simple compression and relaxation of a virtual spring represents an efficient method for generating energy [3]. The produced energy equals the difference of integrals under the curves representing the compression and relaxation of the virtual spring—Fig. 8.1b [4]. From the curves that illustrate conditions for different virtual wall stiffness values can be noticed that energy leakage increases with higher virtual wall stiffness values ($k_2 > k_1$).

A less pronounced effect of implementation of a virtual wall in discrete time is the result of asynchronous transition through the wall surface relative to the sampling time of the controller. Due to discrete sampling, the change in the reaction force does not occur until the next sampling interval after the transition of the wall surface as shown in Fig. 8.1. The time interval denoted by Δt_a indicates the delay in activation of the virtual wall model. The interval denoted by Δt_b indicates the delay in deactivation of the wall model. The result is that the wall model is activated while the spring is already partially compressed, which means an increase in energy since the spring stores energy without a mechanical work being performed. During retraction from the wall, the first sampled free space position will be already outside the wall. Since the wall model is not immediately switched off, the force computed for the last sampled position inside the wall, pushes the haptic display away from the wall, when the display is already in a free space. The result of the asynchronous wall transitions is a net increase in energy that increases the possibility of unstable behavior (chatter) of the haptic interface when in contact with stiff virtual objects.

Tendency to chatter also depends on the characteristics of the human operator, especially on the mechanical impedance of the human arm at the contact point between the human and the haptic display. Thus, when analyzing haptic systems, it is necessary to take into account the interaction of two mechanical systems—a manipulator and a human limb, where both systems are independently controlled. Impedance of the human limb at the contact point with the haptic display can be adapted through muscle activation and changes of the arm and finger postures. Consequently it is possible to initiate and maintain chattering behavior of the haptic interface in certain postures.

8.2 Two-Port Model of Haptic Interaction

As already discussed, haptic interfaces have two main functions. Their first function is measurement of positions and forces (and their time derivatives) of the user, while the second function is display of forces and positions (and their space distributions) resulting from haptic interactions in a virtual or remote (teleoperation) environment. Most often position (velocity) is an input signal to the haptic system and the resulting forces are then displayed to the user—impedance control. On the contrary, in the case of an admittance controller, the haptic interface measures forces applied by the user and displays back positions or velocities to the user.

Linear circuit theory allows analysis of such interactions between the user and virtual or remote (teleoperated) environments. It enables modeling of exchange of energy between the display and the user. The theory, which can be applied to electrical as well as mechanical energy transfers, is appropriate for modeling and control of systems where both types of energy are present simultaneously.

In the case of a haptic display the use of mechanical quantities is more appropriate. Mechanical force is an analogous quantity to voltage and velocity is an analogous quantity to current. With the use of mechanical variables it is possible to schematically represent the interaction between the user and the haptic interface as shown in Fig. 8.2, where F indicates force, v indicates velocity and Z generalized mechanical impedance (inertia, static and viscous damping, elasticity as a function of velocity v and its time derivatives and integrals). Quantities $F_h^\star$ and $F_e^\star$ represent active forces generated by the user and the environment, Z_h and Z_e are impedances of the user and the environment, v_h and F_h are velocity and force at the point of interaction with the haptic display. v_e^* and F_e^* are velocity and force at the connection between the

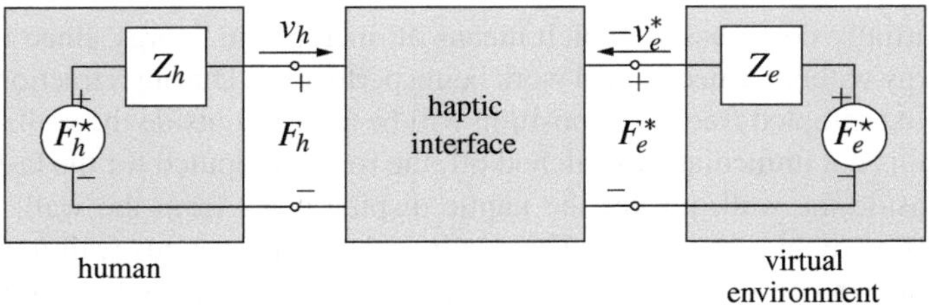

Fig. 8.2 Two-port model of haptic interaction

virtual environment and the haptic interface. A star in the superscript indicates that the variable is of a discrete type.

Two-port models originate from linear circuit theory and allow analysis of stability and performance of bidirectional teleoperation [5, 6]. Two-port models can also be used for the analysis of haptic interactions and characterization of exchange of energy between the user and the virtual environment [7]. A general two-port system is a *black box*, that defines relations between forces (F_h, F_e^*) and velocities (v_h, $-v_e^*$) on two pairs of input terminals.

The dependency between forces and velocities is defined by the *immittance matrix* **P**. A transformation

$$\mathbf{y} = \mathbf{P}\mathbf{u} = \begin{bmatrix} p_{11} & p_{12} \\ p_{21} & p_{22} \end{bmatrix} \mathbf{u} \tag{8.1}$$

defines the *immittance mapping*, if

$$\mathbf{y}^T\mathbf{u} = F_h v_h + F_e^*(-v_e^*). \tag{8.2}$$

A linear two-port system is characterized by a system of two linear equations that relate variables of both input ports: two variables are independent and two variables are dependent. In general, the first two variables represent excitation, while the other two variables represent the system response. Four different immittance matrices are relevant for characterization of haptic interactions: impedance matrix **Z**, admittance matrix **Y**, hybrid matrix **H** and alternative hybrid matrix **G**. Immittance matrices and their components are frequency dependent functions.

An impedance relation is defined as

$$\begin{bmatrix} F_h \\ F_e^* \end{bmatrix} = \mathbf{y}_{\mathscr{Z}} := \mathbf{Z}\mathbf{u}_{\mathscr{Z}} = \begin{bmatrix} z_{11} & z_{12} \\ z_{21} & z_{22} \end{bmatrix} \begin{bmatrix} v_h \\ -v_e^* \end{bmatrix}, \tag{8.3}$$

admittance relation as

$$\begin{bmatrix} v_h \\ -v_e^* \end{bmatrix} = \mathbf{y}_{\mathscr{Y}} := \mathbf{Y}\mathbf{u}_{\mathscr{Y}} = \begin{bmatrix} y_{11} & y_{12} \\ y_{21} & y_{22} \end{bmatrix} \begin{bmatrix} F_h \\ F_e^* \end{bmatrix}, \tag{8.4}$$

hybrid relation as

$$\begin{bmatrix} F_h \\ -v_e^* \end{bmatrix} = \mathbf{y}_{\mathscr{H}} := \mathbf{H}\mathbf{u}_{\mathscr{H}} = \begin{bmatrix} h_{11} & h_{12} \\ h_{21} & h_{22} \end{bmatrix} \begin{bmatrix} v_h \\ F_e^* \end{bmatrix} \quad (8.5)$$

and alternative hybrid relation as

$$\begin{bmatrix} v_h \\ F_e^* \end{bmatrix} = \mathbf{y}_{\mathscr{G}} := \mathbf{G}\mathbf{u}_{\mathscr{G}} = \begin{bmatrix} g_{11} & g_{12} \\ g_{21} & g_{22} \end{bmatrix} \begin{bmatrix} F_h \\ -v_e^* \end{bmatrix}. \quad (8.6)$$

For mappings (8.3)–(8.6) the following equalities hold:

$$\mathbf{y}_{\mathscr{Z}}^T \mathbf{u}_{\mathscr{Z}} = \mathbf{y}_{\mathscr{Y}}^T \mathbf{u}_{\mathscr{Y}} = \mathbf{y}_{\mathscr{H}}^T \mathbf{u}_{\mathscr{H}} = \mathbf{y}_{\mathscr{G}}^T \mathbf{u}_{\mathscr{G}} = F_h v_h + F_e^*(-v_e^*). \quad (8.7)$$

8.3 Stability and Passivity of Haptic Interaction

As shown in Fig. 8.2, a haptic interface can be represented as a two-port network that defines the exchange of energy between the operator (F_h, v_h) and the virtual environment $(F_e^*, -v_e^*)$ [8].

The stability of a two-port depends on its terminal immittances. Immittance characteristics of the operator can be represented either with the impedance Z_h or admittance Y_h. The same applies also for the immittance characteristics of the virtual environment, with impedance defined as Z_e or admittance defined as Y_e.

Definition 8.1 A continuous (discrete) linear two-port network with given terminal immittances is stable if and only if the corresponding characteristic equation has no roots in the right half s-plane (outside of the unit circle in the z-plane) and only simple roots on the imaginary axis (on the unit circle in the z-plane) [7].

A haptic interface must be designed with the assumption that the operator and the virtual environment behave as *a priori* unknown and time variable systems. Most probably the virtual environment is also nonlinear. A stability problem therefore needs to be addressed in conditions of unknown terminal immittances.

The stability concept is often based on the assumption of passive behavior of the operator and the virtual environment. Human limb impedance is adaptive and time dependent. Even though muscles and neural networks controlling the movement are active systems, the human arm nevertheless displays passive characteristics. Studies confirm that the human operator predominantly behaves passively in the frequency range relevant for haptic interaction [9, 10]. A similar requirement cannot be imposed on the virtual environment. Though simulation of passive physical systems (mass, damping, spring) needs to comply with the laws of physics, formulation of numerical integration methods that obey these laws is difficult, if not impossible (virtual spring problem).

However, if it is possible to prove that a two-port network is passive, then the system will be stable if the two-port network is coupled to an arbitrary network that is also passive, which is a sufficient condition for stability of a haptic interface [11–15].

Definition 8.2 A two-port network is passive if and only if the immittance mapping $\mathbf{y} = \mathbf{P}\mathbf{u}$ satisfies the following criterion:

$$\int_0^t \mathbf{y}^T(\tau)\mathbf{u}(\tau)d\tau \geq 0 \qquad \forall t \geq 0 \tag{8.8}$$

for all admissible forces (F_h, F_e^*) and velocities $(v_h, -v_e^*)$ [7]. If only linear two-port networks are considered, passivity criteria can be defined as

$$\begin{aligned} &\Re e\{p_{11}\} \geq 0, \quad \Re e\{p_{22}\} \geq 0 \\ &\Re e\{p_{11}\}\Re e\{p_{22}\} - \left|\frac{p_{21}+p_{12}}{2}\right|^2 \geq 0, \qquad \forall\omega \geq 0. \end{aligned} \tag{8.9}$$

The requirement that the haptic interface behaves passively can be conservative, since it allows arbitrary passive one-port terminations. A less conservative way of solving the stability problem is based on enforcing the *unconditional stability* criterion.

Definition 8.3 A linear two-port is *unconditionally stable* if and only if there exists no set of passive terminating one-port immittances for which the system is unstable. A passive system is always unconditionally stable. On the other hand, an unconditionally stable system is not necessarily passive [7].

For linear two-port networks, Llewellyn's stability criterion defines the necessary and sufficient conditions for unconditional stability,

$$\begin{aligned} &\Re e\{p_{11}\} \geq 0 \\ &2\Re e\{p_{11}\}\Re e\{p_{22}\} \geq |p_{21}p_{12}| + \Re e\{p_{21}p_{12}\} \qquad \forall\omega \geq 0. \end{aligned} \tag{8.10}$$

Conditions (8.10) imply $\Re e\{p_{22}\} \geq 0$.

Llewellyn's criterion can be applied to any form of immittance. If conditions (8.10) are satisfied for one form of immittance, they are also satisfied for the other three immittance forms.

For analysis of haptic interaction, unconditional stability means that the haptic interface will be stable for any set of passive operators and virtual environments. In other words, a haptic interface will remain stable regardless if the operator holds it firmly or releases it completely. At the same time the virtual environment may simulate a free space or an infinitely stiff object. Llewellyn's criterion requires at least locally linearized and time invariant system.

8.4 Haptic Interface Transparency and Z-Width

Quality of a haptic interface can be characterized with its transparency, a property that defines how forces and velocities are transmitted between the user and the virtual environment. A haptic interface with an ideal transparency can be described with a hybrid mapping

$$\begin{bmatrix} F_h \\ -v_e^* \end{bmatrix} = \begin{bmatrix} 0 & 1 \\ -1 & 0 \end{bmatrix} \begin{bmatrix} v_h \\ F_e^* \end{bmatrix}. \tag{8.11}$$

A haptic interface should be as transparent as possible, meaning that it must convey actuator forces to the user consistently when in contact with a virtual environment and should not display motion when there is no contact with virtual objects. Forces applied on the operator are generally smaller than the forces generated by the display actuators due to frictional and inertial losses in the mechanism. For small velocities the viscous damping is negligible and static friction prevails. For high movement velocities the viscous damping prevails over other components. Inertial effects are more pronounced during sudden velocity changes, as, for example, in the initial phase of a contact with a stiff virtual object.

Friction, inertia and other dynamic components prevent the haptic interface from behaving as an ideal device. However, this does not necessarily mean a degraded quality of haptic interaction. Namely, human sensory-motor capabilities prevent the user from discriminating between an ideal and close-to-ideal haptic interface, if discrepancies are not too large.

A different approach to evaluation of a haptic interface is introduced through the concept known as *Z-width* [16]. This is defined as a range of impedances that can be rendered by a haptic interface and presented to the operator without compromising system stability. The range is limited by frequency dependent lower and upper bounds Z_{min} and Z_{max}. An ideal haptic interface enables simulation of a free motion without inertial and frictional forces as well as infinitely stiff and massive objects. Perfect transparency as well as unlimited range of impedances are not achievable goals.

Relations between user movements (displacements or velocities) and the resulting force defines the virtual impedance of a haptic interface. The larger the specter of forces that the haptic interface is able to generate, the larger is the range of virtual impedances. On the other hand, the larger is the space of velocities that a position controlled haptic interface can reproduce, the larger is the range of virtual admittances.

A haptic interface can only render a limited range of impedances, resulting in a limited Z-width. Real-time simulation behaves as a controller for a haptic interface and range of rendered impedances is determined by the stability criteria of the controller. Thus, a haptic interface cannot display arbitrary low or high impedances without compromising stability, a condition that is critical from the operator's perspective.

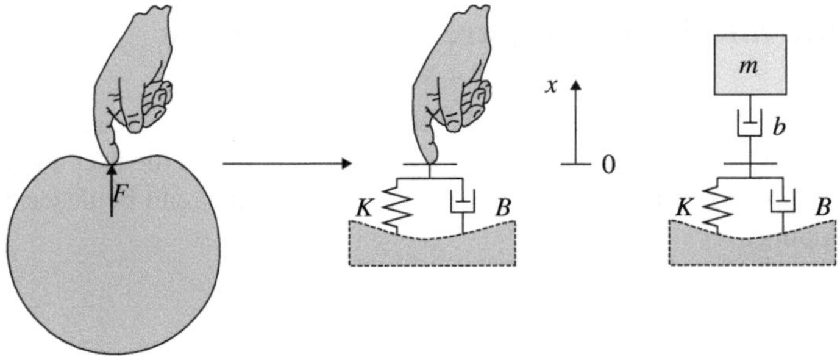

Fig. 8.3 A model of a haptic interface with mass m and viscous damping b in contact with an object with stiffness K and viscous damping B

Next we consider effects of sampling time and intrinsic dynamics of a haptic display on the range of simulated impedances. The analysis will be based on a haptic display with a single degree of freedom. The concept is illustrated in Fig. 8.3.

Conditions for passivity of system (Fig. 8.3) will be analyzed. The following theorem will be used [17]:

Theorem 8.1 *A necessary and sufficient condition for passivity of a model of a haptic interface shown in Fig. 8.4 is*

$$b > \frac{T}{2}\frac{1}{1-\cos(\omega T)}\Re e\{(1-e^{-j\omega T})Z_e(e^{j\omega T})\} \quad \text{for} \quad 0 \le \omega \le \omega_N, \tag{8.12}$$

where b represents intrinsic viscous damping of a haptic display, T is a sampling time, $Z_e(z)$ is the transfer function representing a virtual environment and $\omega_N = \pi/T$ is the Nyquist frequency.

We will consider a special case, a contact with a virtual wall. The analysis will be based on the concept of the virtual wall simulated using a parallel connection of a spring and a damper with unilateral constraint (Figs. 8.3 and 8.4). A velocity estimate is computed from differentiation of a position signal leading to the transfer function

$$Z_e(z) = K + B\frac{z-1}{Tz}, \tag{8.13}$$

where $K > 0$ represents virtual stiffness and B is a coefficient of virtual damping (in general B can be positive or negative). The stability criterion can be obtained by inserting the transfer function (8.13) into the inequality (8.12)

$$\begin{aligned} b &> \frac{T}{2}\frac{1}{1-\cos(\omega T)}\Re e\{(1-e^{-j\omega T})(K + B\frac{e^{j\omega T}-1}{Te^{j\omega T}})\} \\ &= \frac{T}{2}\frac{1}{1-\cos(\omega T)}\left[K\Re e\{(1-e^{-j\omega T})\} + \frac{B}{T}\Re e\{(1-e^{-j\omega T})^2\}\right] \end{aligned}$$

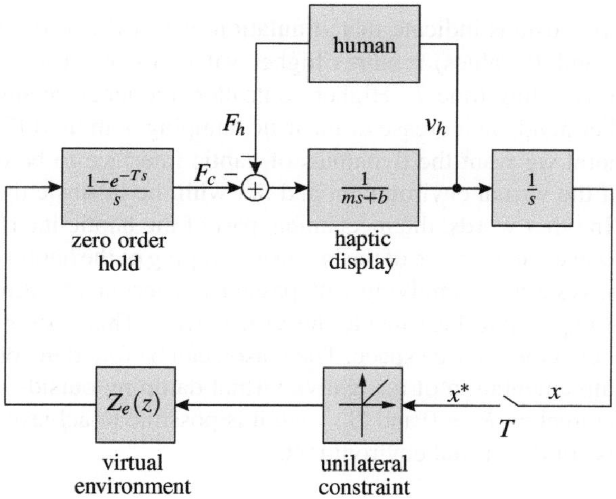

Fig. 8.4 A model of a haptic interface with one degree of freedom. Adapted from [17]

$$
\begin{aligned}
&= \frac{T}{2}\frac{1}{1-\cos(\omega T)}\Big[K\Re e\{1-\cos(\omega T)+j\sin(\omega T)\} \\
&\qquad + \frac{B}{T}\Re e\{(1-\cos(\omega T)+j\sin(\omega T))^2\}\Big] \\
&= \frac{T}{2}\frac{1}{1-\cos(\omega T)}\Big[K(1-\cos(\omega T)) - \frac{2B}{T}(1-\cos(\omega T))\cos(\omega T)\Big].
\end{aligned}
\tag{8.14}
$$

Condition (8.14) leads to

$$b > \frac{KT}{2} - B\cos(\omega T) \quad \text{for} \quad 0 \le \omega \le \omega_N. \tag{8.15}$$

With positive B, the worst conditions occur at $\omega = \omega_N$ (Nyquist frequency), when $\cos(\omega T) = -1$ and consequently

$$b > \frac{KT}{2} + B. \tag{8.16}$$

Analysis of inequalities (8.15) and (8.16) leads to the following conclusions [17]:

- in order to guarantee system passivity, a physical element with damping characteristics is required (intrinsic damping of a haptic display),
- with constant values of b and B the highest achievable virtual stiffness is proportional to the controller sampling frequency,
- in the frequency range below the half of the Nyquist frequency, the virtual damping helps to stabilize the system, while above that frequency the larger B values may lead to violation of the system passivity condition.

The above conclusions indicate that simulation of a stiff and dissipative virtual wall (higher K and B values) requires higher values of intrinsic damping b and lower values of sampling time T. Higher controller frequency seems a reasonable goal. On the other hand, an increase of intrinsic damping is more difficult to justify. Namely, in general we want the dynamics of haptic interface to be determined by the dynamics of the virtual environment and not with the intrinsic dynamics of the haptic display. In other words, the mechanical part of the haptic interface should be transparent. However, an increase of the intrinsic damping of the haptic display assists the discrete time system in complying with passivity criterion, resulting in improved behavior of the haptic interface inside the virtual wall. This does not necessarily deteriorate the behavior in a free space. The reason can be found in condition (8.16), which allows implementation of a negative virtual damping outside the wall. With the choice of parameters $K = 0$ and $B = -b$ it is possible to achieve zero damping and zero stiffness of the virtual environment.

8.5 Virtual Coupling

Stability and performance of haptic interfaces are influenced by different factors. These include device dynamics, sampling effects and in the case of admittance displays also the dynamics of the position control loop.

A simple one degree-of-freedom haptic display, modeled as a mass m and damping b, is shown in Fig. 8.5. Dynamic model of the haptic display is defined by the differential equation

$$m\dot{v}_d + bv_d = F_h - F_d, \quad v_d = v_h, \tag{8.17}$$

where v_h is the user's velocity at the contact point with the display, v_d is the velocity of the display measured at the point of actuation, F_h is the force perceived/produced by the user at the point of contact and F_d is the force produced/measured by the haptic display at the point of actuation.

Laplace transform of (8.17) yields

$$F_h = (ms + b)v_d + F_d = (ms + b)v_h + F_d. \tag{8.18}$$

8.5.1 Impedance Display

Impedance based haptic interaction generates force against the user as a response to measured displacement. From the dynamic model (8.18) it is possible to characterize the impedance display as a hybrid transformation

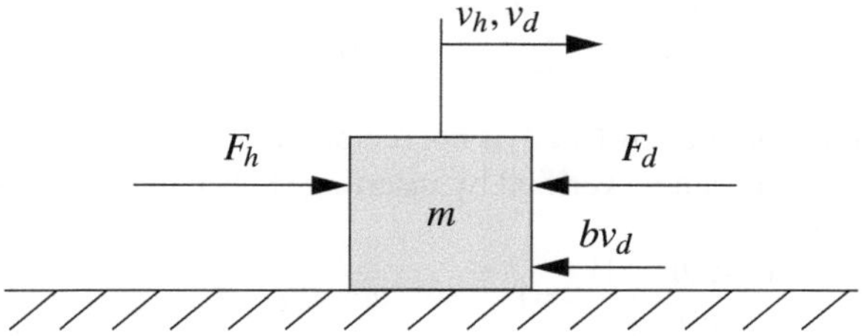

Fig. 8.5 A one-degree-of-freedom haptic display. The user's velocity at the contact point with the display is defined as v_h. Velocity of the display measured at the point of actuation is v_d. Force F_h is detected/produced by the user at the point of contact. Force F_d is produced/measured by the haptic display at the point of actuation

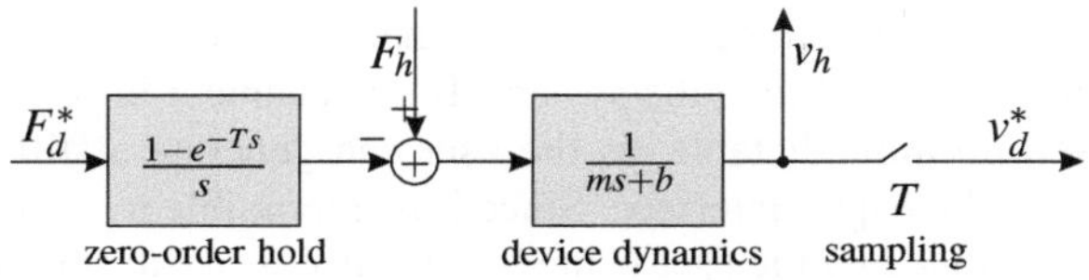

Fig. 8.6 Discretized impedance display with zero-order hold on the signal F_d^* and sampling of velocity v_d^*. Adapted from [7]

$$\begin{bmatrix} F_h \\ -v_d \end{bmatrix} = \begin{bmatrix} ms+b & 1 \\ -1 & 0 \end{bmatrix} \begin{bmatrix} v_h \\ F_d \end{bmatrix}. \tag{8.19}$$

A typical causal structure for haptic interaction is based on an impedance display/impedance environment. The simplest approach is defined by relations $v_d = v_e$ and $F_d = F_e$, where force and velocity at the actuation point of the haptic display are defined by the output of the virtual environment. The resulting immittance matrix is reciprocal and complies with the passivity criteria (8.9) and unconditional stability criteria (8.10). However, this does not mean that the haptic interface is passive and unconditionally stable. In order to properly address these issues, sample and hold effects need to be considered. Figure 8.6 shows a discrete implementation of a haptic interface with the zero-order hold element at the force input port and sampling of the device velocity signal. Open-loop device dynamics is discretized using the Tustin method, which preserves passivity of the impedance function

$$Z_{d_I}(z) = (ms+b)|_{s \to (\frac{2}{T})(\frac{z-1}{z+1})}. \tag{8.20}$$

Zero-order hold and sampling elements can be combined into a transfer function

$$ZOH(z) = \frac{1}{2}\frac{z+1}{z}. \tag{8.21}$$

A discrete hybrid matrix of the impedance display can be written as

$$\begin{bmatrix} F_h \\ -v_d^* \end{bmatrix} = \begin{bmatrix} Z_{d_I}(z)\ ZOH(z) \\ -1 \quad 0 \end{bmatrix} \begin{bmatrix} v_h \\ F_d^* \end{bmatrix}. \tag{8.22}$$

Note that the model defined by (8.22) is not reciprocal. The unconditional stability of the impedance display can be verified by inserting (8.22) into (8.10), which leads to

$$\frac{\Re e\{ZOH(z)\}}{|ZOH(z)|} \geq 1 \quad \Leftrightarrow \quad \cos(\angle ZOH(z)) \geq 1. \tag{8.23}$$

Since the phase shift of the zero-order hold element (8.21) is larger than zero for $0 < \omega < 2\pi/T$, the criterion for unconditional stability is never satisfied.

We may conclude that with equalities $v_d^* = v_e^*$ and $F_d^* = F_e^*$, the haptic interface is never unconditionally stable. This does not mean that the haptic simulation will necessarily be unstable, but there exist a combination of a passive user and virtual environment that will destabilize the system. Biomechanical impedance characteristics of the user are unpredictable, as the user can generate a stiff grasp or may completely release the haptic interface. Since it is impossible to guarantee stability, if stability is conditioned by the user behavior, it is necessary to stabilize the system for any level of user's interaction with the interface, where the only condition is that the user behaves passively.

Since for complex virtual environments it is not easy to guarantee unconditional stability, the problem can be solved by the introduction of a virtual coupling element that guarantees stability of the haptic interface coupled to any virtual environment and any user.

The haptic interface can be redefined as a series connection of a haptic display and a virtual coupling as shown in Fig. 8.7. The goal is to define such a virtual coupling that will guarantee unconditional stability of the system [7, 18]. The virtual coupling can have an arbitrary form. However, physically based virtual coupling is implemented as a spring-damper model with stiffness k_c and damping b_c that connects the haptic display to the virtual environment. Figure 8.8 shows the mechanical analogy of such virtual coupling. When the simulated environment is infinitely stiff, the stiffness perceived by the user is finite and determined by the stiffness of the virtual coupling. Optimal tradeoff between stability and performance is achieved at the highest coupling stiffness that still guarantees unconditional stability of the combined two-port network.

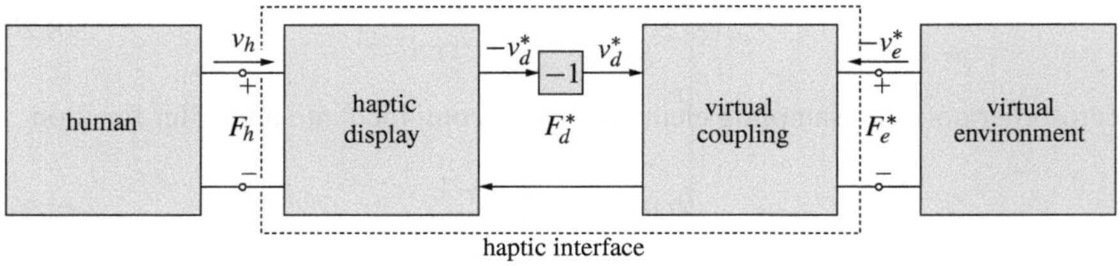

Fig. 8.7 Haptic interface that integrates haptic display and virtual coupling. Adapted from [7]

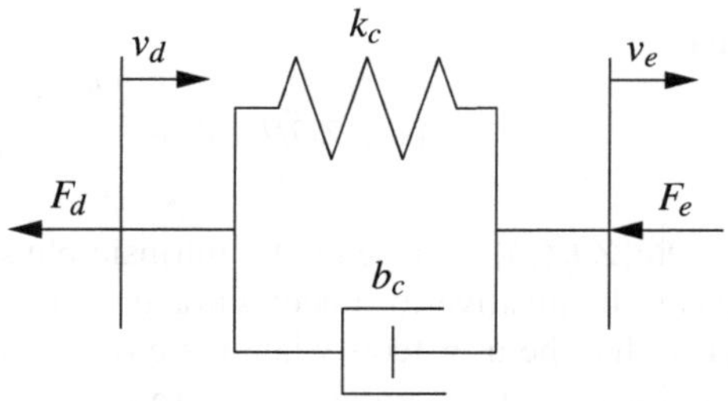

Fig. 8.8 Mechanical analogy of spring-damper based virtual coupling

Discretization of the virtual coupling can be performed as

$$Z_{c_I}(z) = \left(b_c + \frac{k_c}{s}\right)\Big|_{s\to\frac{z-1}{Tz}} . \tag{8.24}$$

A hybrid mapping of the virtual coupling is then

$$\begin{bmatrix} F_d^* \\ -v_e^* \end{bmatrix} = \begin{bmatrix} 0 & 1 \\ -1 & \frac{1}{Z_{c_I}(z)} \end{bmatrix} \begin{bmatrix} v_d^* \\ F_e^* \end{bmatrix} . \tag{8.25}$$

A hybrid mapping characterizing the haptic interface model is a series connection of an impedance display and a virtual coupling. Considering transformations (8.22) and (8.25) we obtain

$$\begin{aligned} &F_h = Z_{d_I}(z)v_h + ZOH(z)F_d^* \\ &F_d^* = F_e^* \quad \Rightarrow \quad F_h = Z_{d_I}(z)v_h + ZOH(z)F_e^* \end{aligned} \tag{8.26}$$

and

$$\begin{aligned} &-v_e^* = -v_d^* + \frac{1}{Z_{c_I}(z)}F_e^* \\ &-v_d^* = -v_h \quad \Rightarrow \quad -v_e^* = -v_h + \frac{1}{Z_{c_I}(z)}F_e^* . \end{aligned} \tag{8.27}$$

Thus, the combined transformation results in

$$\begin{bmatrix} F_h \\ -v_e^* \end{bmatrix} = \begin{bmatrix} Z_{d_I}(z) & ZOH(z) \\ -1 & \frac{1}{Z_{c_I}(z)} \end{bmatrix} \begin{bmatrix} v_h \\ F_e^* \end{bmatrix} . \tag{8.28}$$

By introducing the criterion (8.10), relations that define unconditional stability criteria are

$$\Re e\{Z_{d_I}(z)\} \geq 0, \quad \Re e\left\{\frac{1}{Z_{c_I}(z)}\right\} \geq 0 \tag{8.29}$$

and

$$\cos(\angle ZOH(z)) + \frac{2\Re e\{Z_{d_I}(z)\}\Re e\left\{\frac{1}{Z_{c_I}(z)}\right\}}{|ZOH(z)|} \geq 1. \tag{8.30}$$

$\Re e\{Z_{d_I}(z)\}$ represents the intrinsic physical damping of the haptic display and has to be positive for unconditional stability. $\Re e\{1/Z_{c_I}(z)\}$ represents the compliance that the user feels when the environment simulates a stiff restriction. At least minimal positive compliance is required for satisfying the unconditional stability criteria. Condition (8.30) implies that by increasing the intrinsic damping of the haptic display, it is possible to increase the maximal values of impedances (limited by the virtual coupling impedance $Z_{c_I}(z)$) that can be rendered using a haptic interface without violating the unconditional stability criteria. $\angle ZOH(z)$ means a phase shift due to sampling effects. With a lower sampling frequency, the phase shift increases, requiring either an increase of the intrinsic damping of the device or an increase of the compliance of the virtual coupling in order to guarantee unconditional stability of the system.

From (8.30) we obtain the following criterion for unconditional stability:

$$\Re e\left\{\frac{1}{Z_{c_I}(z)}\right\} \geq \frac{1-\cos(\angle ZOH(z))}{2\Re e\{Z_{d_I}(z)\}}|ZOH(z)| \tag{8.31}$$

that enables computation of optimal virtual coupling parameters.

A hybrid matrix of the combined haptic display and the virtual coupling (8.28) indicates that the best transparency (8.11) can be obtained by increasing $Z_{c_I}(z)$ to its upper limit. For better performance, a high stiffness and damping of the virtual coupling are required. The best virtual coupling is the one that equalizes the left and the right side of the inequality (8.31). This results in an unconditionally stable system with the lowest possible compliance.

The quality of the haptic interface can be estimated from the lower and the upper bounds of the impedance that can be presented to the user $Z_P(z)$, without violating stability criteria. With the termination of the virtual environment port in (8.28) with $F_e^* = Z_e(z)v_e^*$, we can compute the resulting one-port network impedance function $Z_P(z) = \frac{F_h}{v_h}$. Considering relation (8.28) is

$$-v_e^* = -v_h + \frac{1}{Z_{c_I}(z)}F_e^* = -v_h + \frac{Z_e(z)}{Z_{c_I}(z)}v_e^* \quad \Rightarrow \quad v_e^* = \frac{Z_{c_I}(z)v_h}{Z_{c_I}(z)+Z_e(z)} \tag{8.32}$$

and

$$\begin{aligned} F_h &= Z_{d_I}(z)v_h + ZOH(z)F_e^* = Z_{d_I}(z)v_h + ZOH(z)Z_e(z)v_e^* \\ &= \left(Z_{d_I}(z) + ZOH(z)Z_e(z)\frac{Z_{c_I}(z)}{Z_{c_I}(z)+Z_e(z)}\right)v_h, \end{aligned} \tag{8.33}$$

resulting in

$$Z_P(z) = \frac{F_h}{v_h} = Z_{d_I}(z) + \frac{Z_{c_I}(z)ZOH(z)Z_e(z)}{Z_{c_I}(z) + Z_e(z)}. \quad (8.34)$$

This is the impedance perceived by the user when the impedance of the virtual environment equals $Z_e(z)$. The lower bound of Z-width is defined by $Z_e(z) \to 0$. The minimal impedance that the impedance type haptic interface can simulate is limited by its open-loop intrinsic inertia and damping

$$Z_{min_I}(z) = Z_{d_I}(z). \quad (8.35)$$

The upper bound of Z-width can be estimated with the assumption $Z_e(z) \to \infty$. The resulting maximal impedance is the sum of the intrinsic impedance of the device and the impedance of the virtual coupling

$$Z_{max_I}(z) = Z_{d_I}(z) + Z_{c_I}(z)ZOH(z). \quad (8.36)$$

Functions $Z_{min}(z)$ and $Z_{max}(z)$ define the Z-width of the haptic interface. By combining the two equations we obtain

$$Z_{max_I}(z) = Z_{min_I}(z) + Z_{c_I}(z)ZOH(z). \quad (8.37)$$

In order to increase the Z-width of the haptic interface it is necessary to increase the impedance of the virtual coupling or equivalently reduce its compliance.

8.5.2 Admittance Display

Admittance based haptic interaction generates displacement (velocity) as a response to measured interaction force. Admittance approach requires measurement of interaction force at the point of contact between the user and the haptic display and implementation of a velocity feedback loop with a proportional-integral (PI) controller

$$F_d^* = K_{PI}(z)(v_c^* - v_d^*). \quad (8.38)$$

v_c^* is the desired velocity and F_m^* is the measured force. The admittance display is implemented as shown in Fig. 8.9. The resulting alternative hybrid transformation for a discrete time system is given by

$$\begin{bmatrix} v_h \\ F_m^* \end{bmatrix} = \begin{bmatrix} \frac{1}{Z_{d_A}(z)} & -T(z) \\ 1 & 0 \end{bmatrix} \begin{bmatrix} F_h \\ -v_c^* \end{bmatrix}, \quad (8.39)$$

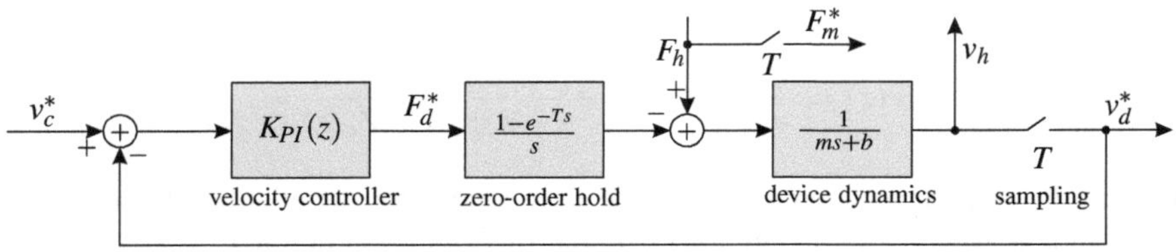

Fig. 8.9 Discretized admittance display with a velocity feedback loop. Adapted from [7]

where

$$T(z) = \frac{ZOH(z)K_{PI}(z)}{Z_{d_I}(z) + ZOH(z)K_{PI}(z)} \tag{8.40}$$

and

$$Z_{d_A}(z) = Z_{d_I}(z) + K_{PI}(z)ZOH(z) \tag{8.41}$$

is the impedance of the admittance display at the point of contact with the user.

A duality between (8.39) and the impedance display model (8.22) may be noted. Forces are mapped to velocities, velocities are mapped to forces, impedance functions are mapped to admittance functions and the force transfer functions are mapped to velocity transfer functions. The duality properties can be useful for the analysis of stability and design of virtual coupling.

By using equalities $v_c^* = v_e^*$ (the desired velocity equals the output of the virtual environment) and $F_m^* = F_e^*$ (the measured force is the input to the virtual environment) and considering Llewellyn's criterion (8.10), it is possible to prove that the admittance display as defined in (8.39) is not unconditionally stable.

The goal is to design a virtual coupling that will guarantee unconditional stability of the combined system as shown in Fig. 8.10 [7]. Since the admittance display is dual to the impedance display, it is possible to presume that the coupling for the admittance display will be dual to the coupling for the impedance display. A mechanical duality of the spring-damper model shown in Fig. 8.8 is a series connection of a mass and damping. Figure 8.11 shows the dual structure. In this case the virtual coupling guarantees a minimal impedance for the virtual environment. Consequently it limits the degree to which the haptic interface can simulate free space. The admittance function of the virtual coupling is

$$Y_{C_A}(z) = \left(\frac{1}{b_c} + \frac{1}{m_c s}\right)\bigg|_{s \to \frac{z-1}{Tz}} . \tag{8.42}$$

With the introduction of a virtual coupling the user perceives viscous damping and inertia of the haptic interface also in conditions, when the virtual environment simulates free space. The best tradeoff between the stability and performance can be obtained with the choice of a virtual coupling with minimal impedance that still guarantees unconditional stability of the two-port network.

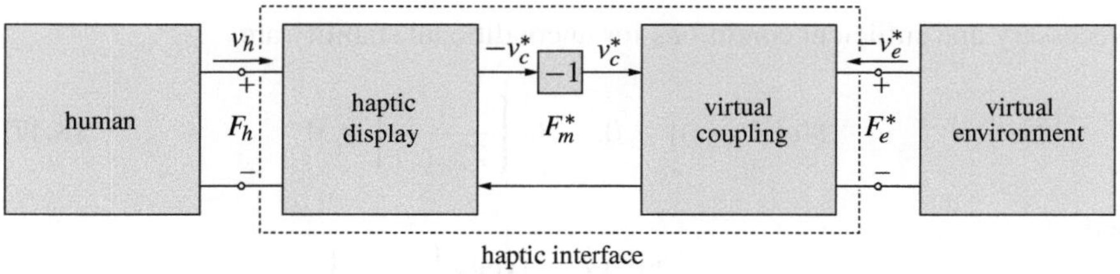

Fig. 8.10 A haptic interface for admittance display includes a haptic display as well as a virtual coupling. Adapted from [7]

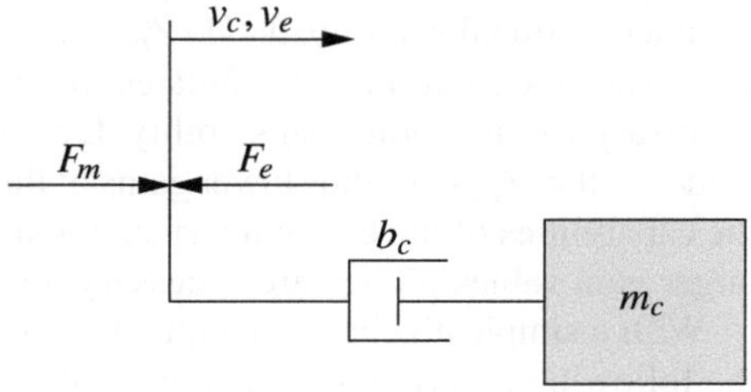

Fig. 8.11 Virtual coupling for an admittance display

An alternative hybrid mapping of the virtual coupling is defined by

$$\begin{bmatrix} v_c^* \\ F_e^* \end{bmatrix} = \begin{bmatrix} 0 & -1 \\ 1 & Z_{c_A}(z) \end{bmatrix} \begin{bmatrix} F_m^* \\ -v_e^* \end{bmatrix}. \tag{8.43}$$

The haptic interface is a series connection of an admittance display and a virtual coupling. Taking into account transformations (8.39) and (8.43) we obtain

$$\begin{aligned} v_h &= \frac{1}{Z_{d_A}(z)} F_h + T(z) v_c^* \\ v_c^* &= v_e^* \quad \Rightarrow \quad v_h = \frac{1}{Z_{d_A}(z)} F_h + T(z) v_e^* \end{aligned} \tag{8.44}$$

and

$$\begin{aligned} F_e^* &= F_m^* - Z_{c_A}(z) v_e^* \\ F_m^* &= F_h \quad \Rightarrow \quad F_e^* = F_h^* - Z_{c_A}(z) v_e^*, \end{aligned} \tag{8.45}$$

resulting in the combined transformation

$$\begin{bmatrix} v_h \\ F_e^* \end{bmatrix} = \begin{bmatrix} \frac{1}{Z_{d_A}(z)} & -T(z) \\ 1 & Z_{c_A}(z) \end{bmatrix} \begin{bmatrix} F_h \\ -v_e^* \end{bmatrix}. \tag{8.46}$$

Necessary and sufficient conditions for unconditional stability are

$$\Re e\{Z_{c_A}(z)\} \geq 0, \quad \Re e\left\{\frac{1}{Z_{d_A}(z)}\right\} \geq 0 \tag{8.47}$$

and

$$\cos(\angle T(z)) + \frac{2\Re e\{Z_{c_A}(z)\}\Re e\left\{\frac{1}{Z_{d_A}(z)}\right\}}{|T(z)|} \geq 1. \tag{8.48}$$

$\Re e\{Z_{c_A}(z)\}$ represents the damping of the virtual coupling that has to be positive for unconditional stability. $\Re e\{1/Z_{d_A}(z)\}$ is the conductance of the admittance display, when the environment simulates stiff constraint. At least minimal compliance is necessary for unconditional stability. Larger values of $\Re e\{1/Z_{d_A}(z)\}$ allow smaller values of $\Re e\{Z_{c_A}(z)\}$, thus lower gains of the velocity feedback loop $K_{PI}(z)$ improve the capabilities of the haptic interface for simulation of free space. At the same time larger gain values $K_{PI}(z)$ are necessary for simulation of stiff constraints.

With a simple algebraic manipulation of inequality (8.48) it is possible to define the following criterion for unconditional stability:

$$\Re e\{Z_{c_A}(z)\} \geq \frac{1 - \cos(\angle T(z))}{2\Re e\left\{\frac{1}{Z_{d_A}(z)}\right\}}|T(z)|. \tag{8.49}$$

The best virtual coupling is the one that only minimally exceeds the lower bound for unconditional stability. The quality of the implementation of the admittance display can be validated by terminating the environment port in (8.46) with $v_e^* = Y_e(z)F_e^*$. By taking into account (8.46) we obtain

$$F_e^* = F_h - Z_{c_A}(z)v_e^* = F_h - Z_{c_A}(z)Y_e(z)F_e^* \quad \Rightarrow \quad F_e^* = \frac{F_h}{1 + Z_{c_A}(z)Y_e(z)} \tag{8.50}$$

and

$$\begin{aligned} v_h &= \frac{1}{Z_{d_A}(z)}F_h + T(z)v_e^* = \frac{1}{Z_{d_A}(z)}F_h + T(z)Y_e(z)F_e^* \\ &= \left(\frac{1}{Z_{d_A}(z)} + \frac{T(z)Y_e(z)}{1 + Z_{c_A}(z)Y_e(z)}\right)F_h. \end{aligned} \tag{8.51}$$

The resulting one-port system function

$$Y_P(z) = \frac{v_h}{F_h} = \frac{1}{Z_{d_A}(z)} + \frac{T(z)Y_e(z)}{1 + Z_{c_A}(z)Y_e(z)} \tag{8.52}$$

represents the admittance perceived by the user, when the admittance of the virtual environment equals $Y_e(z)$. The minimal admittance that can be rendered using the haptic interface can be computed by setting $Y_e(z) \to 0$

$$Y_{min_A}(z) = \frac{1}{Z_{d_A}(z)}. \tag{8.53}$$

The maximal rendered admittance is obtained by setting $Y_e(z) \to \infty$

$$Y_{max_A}(z) = \frac{1}{Z_{d_A}(z)} + \frac{T(z)}{Z_{c_A}(z)}. \tag{8.54}$$

The lower bound of Z-width is defined by the inverse value of the maximal rendered admittance

$$Z_{min_A}(z) = \frac{Z_{c_A}(z)}{T(z) + \frac{Z_{c_A}(z)}{Z_{d_A}(z)}}. \tag{8.55}$$

The upper bound of Z-width is defined by the inverse value of the minimal rendered admittance

$$Z_{max_A}(z) = Z_{d_A}(z) = Z_{d_I}(z) + K_{PI}(z)ZOH(z). \tag{8.56}$$

In order to increase $Z_{max_A}(z)$ and improve the simulation of stiff objects, high $K_{PI}(z)$ gains are required. At the same time the unconditional stability criteria require low values of $K_{PI}(z)$. A tradeoff between stability and performance needs to be found.

The analysis of impedance and admittance displays indicates certain dualities between the two approaches. For unconditional stability intrinsic damping of the haptic display is required. Duality maps this requirement into intrinsic compliance of the admittance display

$$\Re e\{Z_{d_I}(z)\} \geq 0 \quad \leftrightarrow \quad \Re e\left\{\frac{1}{Z_{d_A}(z)}\right\} \geq 0. \tag{8.57}$$

For impedance display a virtual coupling introduces a conductance required for unconditional stability. Duality maps this requirement into impedance of the virtual coupling required for unconditional stability of the admittance display

$$\Re e\left\{\frac{1}{Z_{c_I}(z)}\right\} \geq 0 \quad \leftrightarrow \quad \Re e\{Z_{c_A}(z)\} \geq 0. \tag{8.58}$$

The approximation of the zero-order hold element and the functions related to velocity tracking are dimensionless mappings that have similar effects on stability. $ZOH(z)$ represents the ability of the impedance display for displaying forces to the user when

$v_h = 0$. $T(z)$ represents the ability of the admittance display for displaying velocities of the environment when $F_h = 0$.

8.6 Haptic Interface Stability with Compensation Filter

The stability of a robot force control depends upon the chosen combination of sampling time and controller gain, upon the chosen parameters of compliance (relation between position and force) and also upon stiffness of the contacted environment. Sustained and growing oscillations may appear when a simulated discrete passive virtual environment and a passive human operator are coupled through the haptic interface in a haptic interaction. In admittance type display, the instability will occur when the impedance of the subject is high (e.g. user rigidly grasps the haptic display) and impedance of the virtual environment is low (simulation of free space). The transition from the stable to the unstable state is usually sudden and is accompanied by a pronounced oscillation. Stability therefore needs to be studied with respect to the parameters of the virtual environment and, most importantly, with respect to the stiffness of the coupling between the user and the haptic interface.

8.6.1 Model of Haptic Interaction

A model of haptic interaction is shown in Fig. 8.12. The model consists of a one degree-of-freedom admittance haptic display, a virtual environment and a human operator. The force F_h applied by the human operator is the sum of the voluntary force $F_h^\star$ and force resulting from the biomechanical properties of the limb $G_p(s)v_h$.

The admittance display dynamics usually prevails over the human arm dynamics, thus the human arm dynamics can be modeled as a linearized end-point stiffness $G_p(s) = K$. The assumption is made that the human operator is passive. Force F_h is filtered through the compensation filter $C(s)$ (discussed in detail in Sect. 8.6.2) and the output force F_e is the input to the virtual environment $Y_e(s)$. The force F is the sum of force F_h exerted by the human operator and the controller output force F_c. Position x_e is the desired position and x_h is the actual position. The admittance of the virtual environment Y_e consists of a virtual mass m_e, virtual damping b_e and virtual stiffness k_e in the following form:

$$Y_e(s) = \frac{s}{m_e s^2 + b_e s + k_e}. \tag{8.59}$$

The transfer function $\widehat{Y}_e(s) = \frac{v_h}{F_h}$ is the admittance perceived by the operator as an end-point admittance for the given virtual environment $Y_e(s)$,

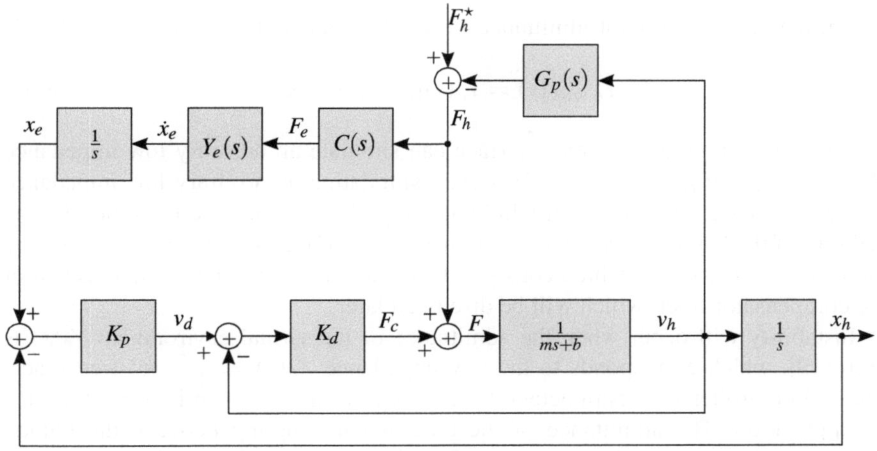

Fig. 8.12 Model of haptic interaction. The model includes a human operator, a virtual environment and an admittance haptic display

$$\widehat{Y}_e(s) = \frac{s + K_d K_p C(s) Y_e(s)}{Z_j(s)s + K_d s + K_d K_p} \tag{8.60}$$

$$= H_i C(s) Y_e(s) + \frac{s}{Z_j(s)s + K_d s + K_d K_p}, \tag{8.61}$$

$$\text{where} \quad H_i(s) = \frac{K_d K_p}{Z_j(s)s + K_d s + K_d K_p}.$$

K_p and K_d are proportional-derivative (PD) controller gains and $Z_j(s)$ represents the haptic display dynamics ($Z_j(s) = ms + b$, m is display mass, b is display damping).

The $H_i(s)C(s)Y_e(s)$ part of (8.61) is the admittance of the virtual environment mapped through $H_i(s)$, which is a model of the one-degree-of-freedom admittance haptic display consisting of a cascade of the PD controller and the manipulator joint dynamics $Z_j(s)$. The $\frac{s}{Z_j(s)s+K_d s+K_d K_p}$ part of (8.61) represents a mechanical coupling between the human operator and the mechanical structure of the haptic display.

The minimum end-point admittance that the haptic interface can simulate is

$$\widehat{Y}_{e,min}(s) = \widehat{Y}_e(s)|_{Y_e \to 0} = \frac{s}{Z_j(s)s + K_d s + K_d K_p}. \tag{8.62}$$

$Y_e \to 0$ is achieved with very high values of the parameters of the virtual environment ($m_e \to \infty$, $b_e \to \infty$ and $k_e \to \infty$; high inertia, highly damped and a very stiff environment, which is the opposite of free space). The minimum admittance of the virtual environment is therefore limited by the impedances of the mechanical coupling between the human operator and the mechanical structure of the haptic display.

The maximum end-point admittance that can be simulated by the haptic interface is

$$\widehat{Y}_{e,max}(s) = \widehat{Y}_e(s)|_{Y_e \to \infty} \to \infty. \tag{8.63}$$

The ideal admittance haptic interface can simulate an arbitrary low impedance $Z_e(s) = \frac{1}{Y_e(s)}$ of the free space. However, simulating an arbitrary low impedance of the free space with an actual admittance type haptic interface is limited by the stability of the haptic interaction. The transparency of free space will be affected by the function $H_i(s)C(s)$, which consists of joint and PD controller dynamics and of the compensator filter, which will be discussed later.

Instability will occur when the admittance of the virtual environment $Y_e(s)$ is high [19], which corresponds to the low impedance $Z_e(s) = \frac{1}{Y_e(s)}$ of free space. The PD controller gain parameters K_d and K_p are high, which is usual for such an application. The admittance of the mechanical coupling between the human operator and the mechanical structure of the haptic display is usually smaller than the $H_i C(s) Y_e(s)$ part of (8.60), thus (8.61) can be simplified for the purpose of studying the stability of the haptic interaction. By adding a representation of the human operator $G_p(s) = K$, the full open-loop transfer function $\widehat{Y}_e(s)G_p(s)$ becomes

$$\widehat{Y}_e(s)G_p(s) = K\frac{K_d K_p C(s) Y_e(s)}{Z_j(s)s + K_d s + K_d K_p} = K H_i(s) C(s) Y_e(s). \tag{8.64}$$

The closed-loop transfer function is

$$x_h = \frac{\widehat{Y}_e(s)}{1 + \widehat{Y}_e(s)G_p(s)} F_h^\star, \tag{8.65}$$

where x_h is the actual end-point position of the haptic interface and $F_h^\star$ is the voluntary force exerted by the human operator. $\widehat{Y}_e(s)$ is in the forward path and the transfer function $G_p(s)$ is in the feedback path.

8.6.2 Design of Compensation Filter

The compensator is used as a filter of the input force exerted by a human operator. The use of a compensation filter results in a haptic control scheme, where a PD controller is used as a feedback controller and a compensation filter is used as a feedforward filter for optimizing the response. The compensation filter is therefore a feedforward part of the control law framework presented in Sect. 8.6.1 and the PD controller is a feedback part of the control law framework. The starting point for deriving the force-filtering compensation filter is the model described with (8.64). $H_i(s)$ is a model of the one-degree-of-freedom admittance haptic display consisting of a cascade of the PD controller and manipulator joint dynamics $Z_j(s)$. In order to achieve stability and good transparency of the haptic interface, $H_i(s)C(s)$ should be close to 1. Since the transfer function $H_i(s)$ is strictly proper (more poles than zeros), $C(s) = \frac{1}{H_i(s)}$ would be a non-causal filter. To preserve the causality of $C(s)$, the following form is adopted:

$$C(s) = \frac{\omega_c^2}{\frac{K_d K_p}{m}} \frac{s^2 + \frac{K_d+b}{m}s + \frac{K_d K_p}{m}}{s^2 + 2\zeta_c\omega_c s + \omega_c^2}$$
$$= \frac{\omega_c^2}{\omega_h^2} \frac{s^2 + 2\zeta_h\omega_h s + \omega_h^2}{s^2 + 2\zeta_c\omega_c s + \omega_c^2}, \tag{8.66}$$

where ω_c determines the new bandwidth of the modeled haptic interface. Parameter ω_c should be chosen to be greater than the time constants of $H_i(s)$, then $H_i(s)C(s) \approx 1$ for $\omega < \omega_c$, which improves the overall transparency of the haptic interface. $H_i(s)C(s)$ takes the form

$$H_i(s)C(s) = \frac{\omega_c^2}{s^2 + 2\zeta_c\omega_c s + \omega_c^2}. \tag{8.67}$$

The minimum and maximum end-point admittance and, hence the Z-width of the haptic interface with the compensation filter, do not change. Figure 8.13 shows a

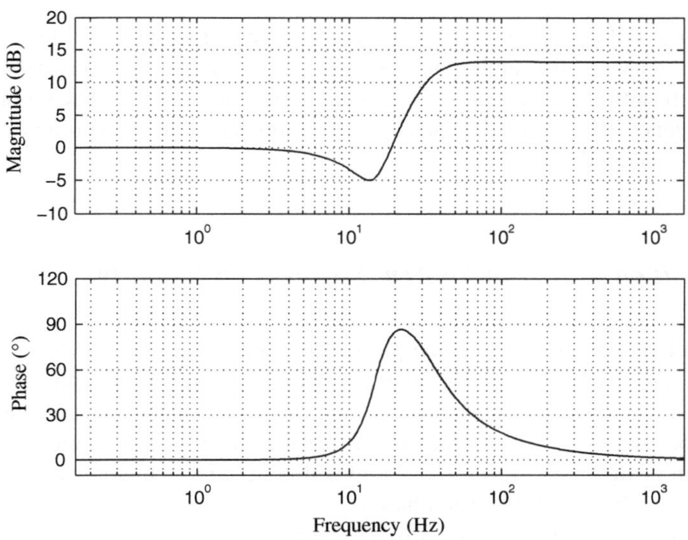

Fig. 8.13 Bode diagram of the force compensation filter. Parameters of the force filter compensator should be chosen such that the positive phase is in the region where phase margin has to be improved

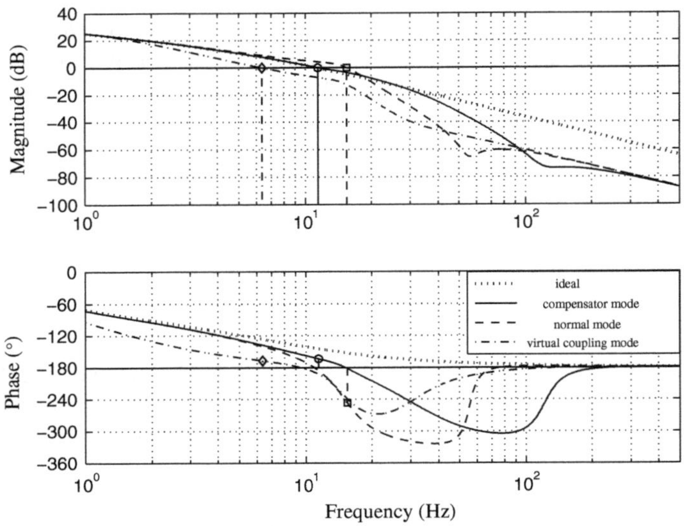

Fig. 8.14 Open loop Bode plots of ideal (*dotted line*), uncompensated (*broken line*), compensated (*solid line*) haptic interface and haptic interface with virtual coupling (*dash-dotted line*). *Circle* indicates gain cross-over frequency of compensated system, *square* represents the gain cross-over frequency of uncompensated system and *diamond* represents the gain cross-over frequency of system with virtual coupling

Bode diagram of the force filter $C(s)$. The positive angle phase of filter $C(s)$ raises the overall curve of the angle phase and increased damping moves unstable poles into the left-half of the complex plane, thus improving stability. The amount of the high-frequency attenuation depends upon the ratio between the denominator and the numerator bandwidths.

8.6.3 Influence of Human Arm Stiffness on Stability

Figure 8.14 shows open loop Bode plots of the haptic interaction for ideal system ($\widehat{Y_e}(s)G_p(s) = KY_e(s)$), for the uncompensated system, for the compensated system and for the system with the virtual coupling. Biomechanical impedance of the human operator $G_p(s) = K$ does not affect the shape of the magnitude plot, but shifts it up or down by $20log_{10}K$. An ideal haptic interface is always stable, with phase always above $-180°$. The phase of an actual system drops below $-180°$ for the uncompensated system, the compensated system and the system with virtual coupling in the high-frequency region due to display and PD controller dynamics, since they introduce additional poles into the system. The added compensation filter damps the resonant peak of the uncompensated system and increases the phase of the overall system in critical frequency region. In addition, the interface transparency is improved, since the magnitude and phase of the compensated system are closer to ideal than of those of the uncompensated case (see Fig. 8.14). Similarly, the virtual coupling damps the oscillating peak and increases the phase of the overall system. However, on the basis of Fig. 8.14, it can be seen that the virtual coupling affects transparency considerably more than the compensation filter.

8.6.4 Compensation Filter and Input/Loop-Shaping Technique

The design proposed in Sect. 8.6.2 has a number of similarities to the input-shaping techniques (IST) applied in industrial robotics for vibration reduction [20] and the loop-shaping method for robust performance design.

The IST scheme utilizes a feedforward controller for suppressing vibrations and a feedback controller for attaining robustness against disturbances or parameter variations. Similarly, the approach shown here uses the compensation filter as a feedforward controller for preshaping force F_h in order to suppress the magnitude and to raise the phase in the resonant frequency region. The PD controller is used as a feedback controller.

The idea behind loop-shaping is to construct an open-loop transfer function in such a manner that the feedback system is internally stable so that it satisfies the robust performance condition. Section 8.6.2 presents the procedure for $C(s)$ selection such that $\widehat{Y}_e(s)$ arbitrarily approximates $Y_e(s)$ by choosing the best performance values of ω_c and ζ_c.

For the best performance, the design of the compensation filter suggests values of ω_c frequency as high as possible. However, this will lead to a high attenuation $C(j\omega)|_{\omega \longrightarrow \infty} = \frac{\omega_c^2}{\omega_h^2}$ at high frequencies, which is not desirable. Tan et al. report that the upper bound of the human force control bandwidth is about $\omega_F = 7$ Hz [21]. Hence, ω_c should at least match the haptic interface bandwidth ω_h and the human force control bandwidth ω_F. However, for improvement of stability the compensation filter bandwidth ω_c should be chosen higher than the approximate value of ω_h to ensure a sufficient stability margin. The utilized compensation filter has a simple design and can be implemented usefully, even if no strict identification of the haptic interface has been made.

A very important property of the described compensation filter is that it is not present in a feedback part of the control law framework, but is included in a feedforward chain. This enables independent design of the compensator and the feedback controller [22]. The described compensation filter can be combined with a number of known feedback algorithms for haptic interface control frameworks.

8.7 Passivity of Haptic Interface

In the previous chapters we introduced concepts of virtual coupling that guarantee stability of haptic interfaces. We assumed passive behavior of the virtual environment. However, discrete time controller implementation may cause parts of the system to behave actively. This active behavior can be detected and appropriate measures can be taken to guarantee passivity of the haptic interface. Therefore, concepts of a *passivity observer* and a *passivity controller* will be introduced [11, 12].

As a starting point, condition (8.8) for passivity of a two-port network will be used. Signs of forces and velocities are defined such that their product is positive when power enters the two-port system. We assume that at time $t = 0$ the energy stored in the system equals $E(0)$ and we rewrite the passivity condition (8.8).

Definition 8.4 The one-port network N with the initial stored energy $E(0)$ is passive if and only if

$$\int_0^t F(\tau)v(\tau)d\tau + E(0) \geq 0, \quad \forall t \geq 0 \tag{8.68}$$

for all admissible forces F and velocities v [12]. Equation (8.68) states that the passive system consumes energy.

Elements of the typical haptic system include virtual environment, virtual coupling, controller of the haptic device, haptic display and a user. Most of the input and output quantities of the haptic system components can be measured in real time. Thus, Eq. (8.68) can also be computed in real time.

8.7.1 Passivity Observer

Quantities that determine the flow of energy are discrete variables. We assume a system, where the sampling frequency is much higher than the dynamics of the haptic device, the user and the virtual environment. Thus, it is possible to assume that the change of force and velocity in a single sampling period is small. In this case it is possible to equip one or more subsystems with the passivity observer

$$E_{obsv}(n) = T \sum_{k=0}^{n} F(k)v(k), \tag{8.69}$$

where T is the sampling time.

If $E_{obsv}(n) \geq 0$ for all n, then the system consumes energy. If at any time $E_{obsv}(n) < 0$, the system generates energy and the quantity of the generated energy equals $-E_{obsv}(n)$.

8.7.2 Passivity Controller

A passivity observer provides information about the quantity of the energy flowing in and out of the system. Therefore, it is possible to design a time dependent element that consumes the excessive energy. Such an element is called a *passivity controller*.

A passivity controller takes the form of a dissipative element connected in series or in parallel configuration as shown in Fig. 8.15. The passivity controller is defined by the relation

$$F = \alpha v, \tag{8.70}$$

where α represents variable damping. Specifically for a series connection the relation is defined as

$$F_1 = F_2 + \alpha v_2 \tag{8.71}$$

and for parallel connection

$$v_2 = v_1 - \frac{F_1}{\alpha}. \tag{8.72}$$

For a series connection of a passivity controller with impedance based virtual environment the value α can be computed by assuming that $v_1(n) = v_2(n)$ is the input and $F_2(n) = F_E(v_2(n))$, where F_E is the output of the virtual environment. The passivity observer is then

$$E_{obsv}(n) = E_{obsv}(n-1) + \left(F_2(n)v_2(n) + \alpha(n-1)v_2(n-1)^2\right)T. \tag{8.73}$$

Based on the output of the passivity observer, the passivity controller can be computed as

$$\alpha(n) = \begin{cases} -\frac{E_{obsv}(n)}{Tv_2(n)^2} & \Leftarrow \quad E_{obsv}(n) < 0 \\ 0 & \Leftarrow \quad E_{obsv}(n) \geq 0. \end{cases} \tag{8.74}$$

Finally, output force is then defined as

$$F_1(n) = F_2(n) + \alpha(n)v_2(n). \tag{8.75}$$

It is possible to prove that the system connected to the passivity controller (8.74), is passive:

$$T\sum_{k=0}^{n} F_1(k)v_1(k) = T\sum_{k=0}^{n} F_2(k)v_2(k) + T\sum_{k=0}^{n} \alpha(k)v_2(k)^2$$

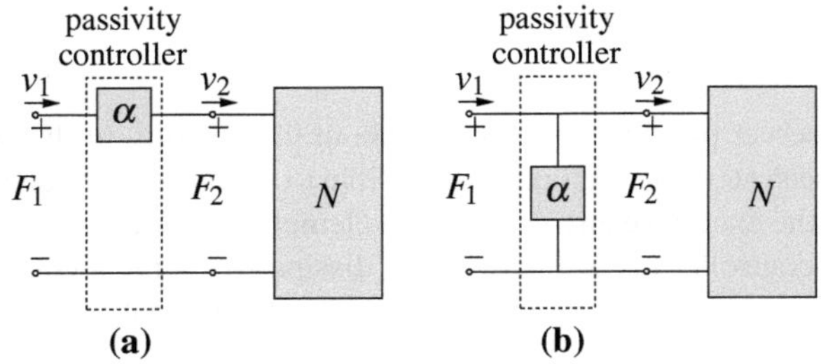

Fig. 8.15 **a** series and **b** parallel connection of a passivity controller for a one-port network. α is an adjustable damping element. The choice of connection depends on the input/output causal structure of the model describing the one-port network. Adapted from [12]

$$= T\sum_{k=0}^{n} F_2(k)v_2(k) + T\sum_{k=0}^{n-1} T\alpha(k)v_2(k)^2 + \alpha(n)v_2(n)^2$$
$$= E_{obsv}(n) + T\alpha(n)v_2(n)^2. \tag{8.76}$$

By inserting Eq. (8.74) we obtain

$$T\sum_{k=0}^{n} F_1(k)v_1(k) \geq 0 \quad \forall n. \tag{8.77}$$

Similarly, the parameter α can be computed for a parallel connection and an admittance type of virtual environment by assuming that $F_1(n) = F_2(n)$ is the input and $v_2(n) = v_E(F_2(n))$, where v_E is the output of the virtual environment. The passivity observer is then

$$E_{obsv}(n) = E_{obsv}(n-1) + T(F_2(n)v_2(n) + \frac{1}{\alpha(n-1)}F_2(n-1)^2). \tag{8.78}$$

Based on the output of the passivity observer, the passivity controller can be computed as

$$\frac{1}{\alpha(n)} = \begin{cases} -\frac{E_{obsv}(n)}{T F_2(n)^2} & \Leftarrow \quad E_{obsv}(n) < 0 \\ 0 & \Leftarrow \quad E_{obsv}(n) \geq 0. \end{cases} \tag{8.79}$$

Finally, output velocity is then defined as

$$v_1(n) = v_2(n) + \frac{1}{\alpha(n)}F_2(n). \tag{8.80}$$

It is possible to prove that the system connected to the passivity controller (8.79), is passive:

$$T\sum_{k=0}^{n} F_1(k)v_1(k) = T\sum_{k=0}^{n} F_2(k)v_2(k) + T\sum_{k=0}^{n-1} \frac{1}{\alpha(k)} F_2(k)^2 + \frac{T}{\alpha(n)} F_2(n)^2$$
$$= E_{obsv}(n) + \frac{T}{\alpha(n)} F_2(n)^2 \tag{8.81}$$

By inserting (8.79) in the above equation we obtain

$$T\sum_{k=0}^{n} F_1(k)v_1(k) \geq 0 \quad \forall n. \tag{8.82}$$

References

1. Mahaparta, S., Žefran, M.: Stable haptic interaction with switched virtual environments. In: Proceedings of IEEE International Conference on Robotics and Automation, pp. 1241–1246 (2003)
2. Colgate, J.E., Grafing, P.E., Stanley, M.C., Schenkel, G.: Implementation of stiff virtual walls in force-reflecting interfaces. In: IEEE Virtual Reality Annual International Symposium, VRAIS 93, pp. 202–208 (1993)
3. Gillespie, R.B., Cutkosky, M.R.: Stable user-specific haptic rendering of the virtual wall. In: Proceedings of the ASME Dynamic Systems and Control Division, pp. 397–406 (1996)
4. Basdogan, C., Srinivasan, M.A.: Handbook of Virtual Environments: Design, Implementation, and Applications, Chap. Haptic Rendering In Virtual Environments, pp. 117–134. Lawrence Erlbaum Associates, New Jersey (2001)
5. Hannaford, B.: A design framework for teleoperators with kinesthetic feedback. IEEE Trans. Robot. Autom. **5**, 426–434 (1989)
6. Raju, G., Verghese, G.C., Sheridan, T.B.: Design issues in 2-port network models of bilateral remote manipulation. In: Proceedings of the IEEE International Conference on Robotics and Automation, pp. 1316–1321 (1989)
7. Addams, R.J., Hannaford, B.: Stable haptic interaction with virtual environments. IEEE Trans. Robot. Autom. **15**, 465–474 (1999)
8. Addams, R.J., Hannaford, B.: Control law design for haptic interfaces to virtual reality. IEEE Trans. Control Syst. Technol. **10**, 3–13 (2002)
9. Hogan, N.: Controlling impedance at the man/machine interface. In: Proceedings of the IEEE International Conference on Robotics and Automation, pp. 1626–1631 (1989)
10. Popescu, F.C.: Physiology and stability criterion for human robot interaction. In: Proceedings of International Conference on, Advanced Robotics, pp. 1456–1461 (2003)
11. Hannaford, B., Ryu, J.H., Kim, Y.S.: Stable control of haptics. In: Proceedings USC Workshop on Haptic Interfaces, Touch in Virtual Environments (2001)
12. Hannaford, B., Ryu, J.H.: Time-domain passivity control of haptic interfaces. IEEE Trans. Robot. Autom. **18**, 1–10 (2002)
13. Stramigioli, S., Secchi, C., van der Schaft, A.J., Fantuzzi, C.: A novel theory for sampled data system passivity. In: Proceedings of IEEE/RSJ International Conference on Intelligent Robots and Systems, pp. 1936–1941 (2002)
14. Mahvash, M., Hayward, V.: Passivity-based high-fidelity haptic rendering of contact. In: Proceedings of the IEEE International Conference on Robotics and Automation (2003)

15. Spong, M.W.: The passivity paradigm in robot control. In: Plenary Lecture at the Chinese Control Conference, Wuxi, China (2004)
16. Colgate, J.E., Stanley, M.C., Brown, J.M.: Issues in the haptic display of tool use. In: Proceedings of the 1995 IEEE/RSJ International Conference on Intelligent Robots and Systems, Pittsburgh, pp. 140–145 (1995)
17. Colgate, J.E., Brown, J.M.: Factors affecting the z-width of a haptic display. In: Proceedings of the IEEE International Conference on Robotics and Automation, pp. 3205–3210 (1994)
18. Lee, M.H., Lee, D.Y.: Nonlinear virtual coupling based on a human dynamics model. In: Proceedings of International Conference on, Advanced Robotics, pp. 1443–1448 (2003)
19. Carignan, C.R., Cleary, K.R.: Closed-loop force control for haptic simulation of virtual environments. Haptics-e **1**, 1–14 (2000)
20. Changa, P.H., Parkb, H.S.: Time-varying input shaping technique applied to vibration reduction of an industrial robot. Control Eng. Pract. **13**(1), 121–130 (2005)
21. Tan, H.Z., Srinivasan, M.A., Eberman, B., Cheng, B.: Human factors for the design of force-reflecting haptic interfaces. Dyn. Syst. Control **55**, 353–359 (1994)
22. Yaesh, I., Shaked, U.: Two-degree-of-freedom h infin;-optimization of multivariable feedback systems. IEEE Trans. Autom. Control **36**(11), 1272–1276 (1991)

Chapter 9
Teleoperation

A teleoperation system enables the human operator to remotely explore and manipulate objects. It usually simplifies the interaction between the human and the robot to a level that allows completion of tasks that neither the human nor the robot could perform alone. A teleoperation system consists of three parts:

- a master device (a haptic interface) that the human operator holds and manipulates,
- a slave device that manipulates objects based on the commands from the master device,
- a controller that couples master and slave devices by transmitting movements and forces between the two devices.

The interaction between the operator and the master device is usually limited to the end-effector of the master device. Similarly, the slave device acts on the environment only with a tool or a gripper mounted on the device end-effector. Kinematics of master and slave devices can be equal, scaled or completely different. However, the number of degrees of freedom of the master device should be equal or larger than the number of active (controlled) degrees of freedom of the slave device. A controller smooths the differences in kinematics of the two devices. An example of a teleoperation system is shown in Fig. 9.1.

Teleoperation combines the best of human and robot characteristics. A slave device has similar functions as an autonomous robot, except that it is controlled in real time via the user manipulation. Robots, even though augmented with artificial intelligence, are not capable of human reasoning and intuition. Even on the level of motor skills, there are no control algorithms currently available that would be capable of complex interactions that include force and position (haptic interactions), something that a human can perform unconsciously during everyday activities.

On the other hand, robots have several advantages compared to humans. They are not susceptible to high temperatures, pressures and radiation. With a specific design of robot mechanisms it is possible to construct devices of microscopic to macroscopic dimensions that are cleaner, stronger, faster and more accurate than any human.

Considering the complementary capabilities of a human and a robot the best teleoperation system is the one that combines the best characteristics of each system.

M. Mihelj and J. Podobnik, *Haptics for Virtual Reality and Teleoperation*, Intelligent Systems, Control and Automation: Science and Engineering 64,
DOI: 10.1007/978-94-007-5718-9_9,

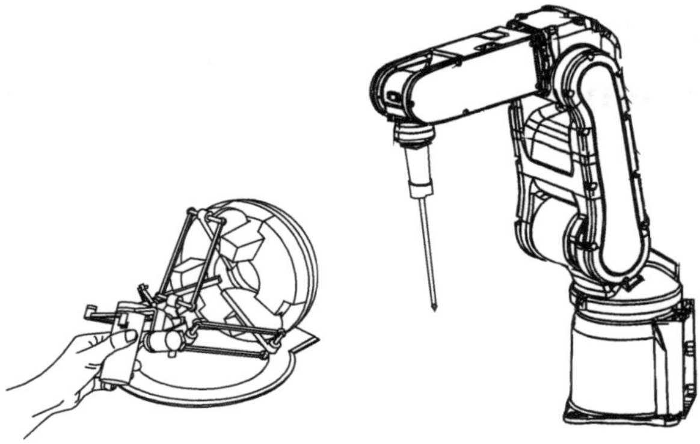

Fig. 9.1 En example of a teleoperation system with the master and slave devices

An ideal teleoperation system guarantees that the trajectory of a tool mounted on a slave device end-effector perfectly matches the operator's movements. Forces acting on the tool are precisely mapped back to the human operator. However, this may result in unstable behavior of the teleoperation system. Namely, the impedance of a master device is usually much lower than the impedance of a slave device, stiffness of the operator's arm is relatively low and the inertia of the human arm is almost negligible compared to the inertia of the robot used as a slave device. Thus, the operator may not absorb all the master device forces resulting in an involuntary arm movement that is transmitted via the master device to the slave device if perfect trajectory tracking is assumed. Thus, the system becomes destabilized due to the differences in dynamics between the slave device and the operator.

It should be noted that the problems related to teleoperation are different from those related to virtual reality. In the case of teleoperation signals from the slave to the master device are limited by the laws of physics, while in the case of a virtual environment simulation this is not necessarily the case. At the same time a haptic interface is only one of the active mechanical components in a teleoperation system, the other one is the slave device.

9.1 Two-Port Model of Teleoperation

A teleoperation system is usually designed with the goal of assisting a human operator in performing manipulation in remote and dangerous environments or executing precision tasks via master-slave robotic devices. Besides stability, which is critical for any control system, transparency is the most important design parameter for implementation of a controller. Transparency, which represents a detailed rendering of a remote environment to the human operator, can be achieved if position and force

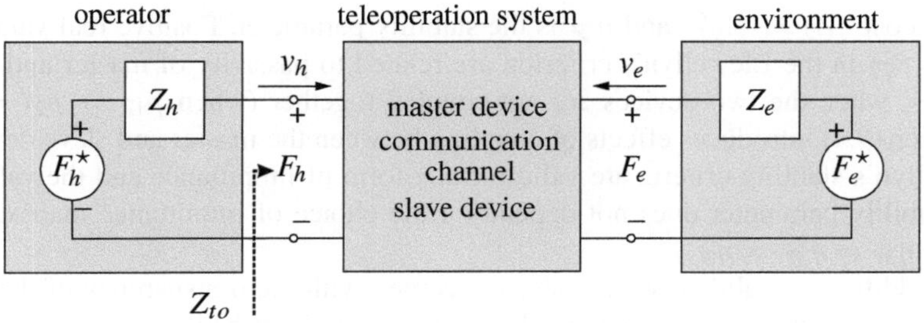

Fig. 9.2 A teleoperation system including the human and the environment dynamics. Adapted from [5]

applied by the slave device faithfully track position and force at the master device [1]. However, perfect transparency cannot be achieved due to delays in data transmission between master and slave devices and variable dynamics of the operator and the system, which significantly affect stability and performance of the teleoperation system [2, 3].

Assume conditions as shown in Fig. 9.2, where a master device, a slave device and a communication channel are modeled as linear, time invariant two-port systems [4]. Assume a contact between the operator and the master device as well as a contact between the slave device and the environment, where the two contacts can be modeled as

$$F_h = F_h{}^\star - Z_h v_h \tag{9.1}$$

$$F_e = F_e{}^\star - Z_e v_e, \tag{9.2}$$

where Z_h, Z_e, v_h, v_e represent impedances and velocities of the master and slave devices, F_h is a force applied by the operator on the master device, F_e is a force applied by the slave device on the environment and $F_h{}^\star$ in $F_e{}^\star$ are active forces generated by the operator and the environment.

For a linear time invariant two-port network, Llewellyn's criterion (Definition 8.3 in Chap. 8) defines the necessary and sufficient conditions for unconditional stability of the teleoperation system. Condition (8.10) can be rewritten as

$$\Re e\{p_{11}\} \geq 0, \quad \Re e\{p_{22}\} \geq 0, \tag{9.3}$$

$$\begin{aligned} \eta_{\mathscr{P}} &= -\frac{\Re e\{p_{21}p_{12}\}}{|p_{21}p_{12}|} + 2\frac{\Re e\{p_{11}\}\Re e\{p_{22}\}}{|p_{21}p_{12}|} \\ &= -\cos(\angle|p_{21}p_{12}|) + 2\frac{\Re e\{p_{11}\}\Re e\{p_{22}\}}{|p_{21}p_{12}|} \geq 1 \quad \forall \omega \geq 0, \end{aligned} \tag{9.4}$$

where $\cos(\angle Z) = \frac{\Re e\{Z\}}{|Z|}$ and $\eta_{\mathscr{P}}$ is the stability parameter. Positive real values of p_{11} in p_{22} in the Llewellyn's criterion are related to passivity of master and slave devices, when the two devices are not coupled together (when $p_{12} = p_{21} = 0$). Equation (9.4) introduces effects of coupling between the master and slave devices. Llewellyn's stability criteria are valid for any form of immittance and the value of the stability parameter does not depend on the choice of immittance matrix, thus $\eta_{\mathscr{Z}} = \eta_{\mathscr{Y}} = \eta_{\mathscr{H}} = \eta_{\mathscr{G}}$.

In addition to stability, we are also concerned with the transparency of the teleoperation system. Transparency can be qualitatively described as a match between the environment impedance and the impedance transferred to the human operator; thus, an ideal transparency can be defined based on Fig. 9.2 as $Z_{to} = \frac{F_h}{v_h}|_{F_e^\star=0} = \frac{F_e}{v_e}|_{F_e^\star=0} = Z_e$, where Z_{to} is the impedance being perceived by the operator. Taking into account Eqs. (9.1), (9.2) and the hybrid immittance transformation (8.5), we can define Z_{to} as a function of hybrid parameters

$$F_h = h_{11}v_h + h_{12}F_e = h_{11}v_h + h_{12}Z_e v_e \tag{9.5}$$

$$-v_e = h_{21}v_h + h_{22}F_e = h_{21}v_h + h_{22}Z_e v_e \tag{9.6}$$

$$v_e = -\frac{h_{21}v_h}{1 + h_{22}Z_e} \tag{9.7}$$

$$Z_{to} = \frac{F_h}{v_h} = h_{11} - h_{12}\frac{h_{21}}{1 + h_{22}Z_e}Z_e = \frac{h_{11} + \Delta h Z_e}{1 + h_{22}Z_e}, \tag{9.8}$$

where $\Delta h = h_{11}h_{22} - h_{12}h_{21}$. If hybrid parameters are not a function of impedances Z_h and Z_e, perfect transparency is achieved when the following conditions are satisfied:

$$\begin{aligned} h_{11} &= h_{22} = 0 \\ h_{12} &= -h_{21} = 1. \end{aligned} \tag{9.9}$$

A perfectly transparent system is marginally stable, as by taking into account conditions (9.9) in Llewellyn's stability criteria we obtain $\Re e\{h_{11}\} = 0$ and stability parameter $\eta_{\mathscr{H}} = 1$. Thus, in order to increase system stability margin, its transparency needs to be reduced. Additionally, delays in the communication channel between the master and the slave devices need to be considered, which further deteriorate stability conditions. Therefore, a compromise between stability and transparency needs to be found. Z-width of a teleoperation system can be determined by taking into account extreme impedance values of the environment. The lowest impedance that can be displayed to the operator is defined by the free movement of the slave device when there is no contact with the environment $Z_e = 0$; the highest impedance is defined, when the slave device movement is locked $Z_e \to \infty$. Taking into account Eq. (9.8) we obtain the boundary values of impedances displayed to the operator:

$$
\begin{aligned}
Z_{to_{min}} &= Z_{to}|_{Z_e=0} = h_{11} \\
Z_{to_{max}} &= Z_{to}|_{Z_e\to\infty} = Z_{to_{min}} + \frac{-h_{12}h_{21}}{h_{22}}.
\end{aligned}
\tag{9.10}
$$

A high quality teleoperation system is characterized by $Z_{to_{min}} \to 0$ and $Z_{to_{max}} \to \infty$. The difference between $Z_{to_{max}}$ and $Z_{to_{min}}$ defines the Z-width of the teleoperation system.

9.2 Teleoperation Systems

In general it is possible to characterize teleoperation systems as impedance or admittance type devices, depending on their behavior as force or velocity generators. Linear time invariant models of impedance/admittance type of master and slave devices can be defined as follows:

- for an impedance type master device

$$Z_m v_h = F_h + F_{cm}, \tag{9.11}$$

  where Z_m describes the dynamics (impedance that is typically low) of the master device and F_{cm} control variable for the master device,
- for an impedance type slave device

$$Z_s v_e = -F_e + F_{cs}, \tag{9.12}$$

  where Z_s describes the dynamics (impedance that is typically low) of the slave device and F_{cs} control variable for the slave device,
- for an admittance type master device

$$Y_m F_h = v_h + v_{cm}, \tag{9.13}$$

  where Y_m describes the dynamics (admittance that is typically low) of the master device and v_{cm} control variable for the master device and
- for admittance type slave device

$$Y_s F_e = -v_e + v_{cs}, \tag{9.14}$$

  where Y_s describes the dynamics (admittance that is typically low) of the slave device and v_{cs} control variable for the slave device.

Based on the system classification listed above, four types of teleoperation systems can be defined that combine different dynamics of master and slave devices [5]. Figure 9.3 shows a block scheme of a teleoperation system, where general four-channel

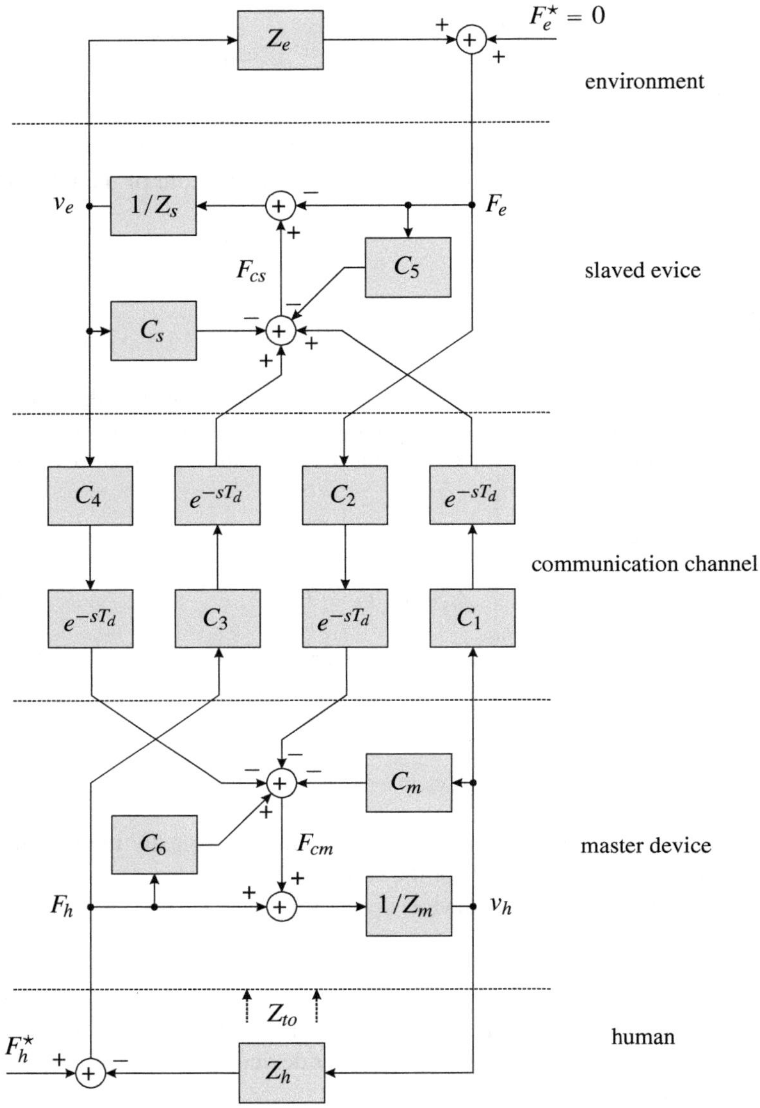

Fig. 9.3 Four-channel control architecture for a teleoperation system. Adapted from [5]

bilateral controllers are used [1, 5] (other four-channel configurations can be found in [5]).

T_d indicates delays inherent to the communication channel, blocks C_i indicate controller transfer functions, Z_h and Z_e represent operator's and environment impedances, while $F_h^\star$ and $F_e^\star$ indicate operator's and environment exogenous forces.

Two types of quantities are generally used for controlling actuators of master and slave devices. The first type is defined by the outputs of blocks C_5 (slave local force controller), C_6 (master local force controller), C_m (master local position controller) and C_s (slave local position controller). These controllers are built around the master and slave devices. The second type is defined by outputs of blocks that define feedforward controllers (blocks $C_1, \ldots, C_4$), that transmit information about forces and velocities between the master and slave devices. The form of the feedforward signal (force or velocity) depends on the type of master and slave devices (impedance or admittance). The output variable of a feedforward compensation for impedance type manipulator is force. Thus C_2 (slave force feedforward controller) and C_3 (master force feedforward controller) are simple gains for scaling measured forces on master and slave devices, while C_1 (master coordinating feedforward controller) and C_4 (slave coordinating feedforward controller) are impedance filters that map measured velocities of master and slave devices into forces.

9.3 Four-Channel Control Architecture

Analysis of a four-channel control architecture will be based on relations in Fig. 9.3. Control variables for master and slave devices F_{cm} and F_{cs} can be determined as

$$F_{cm} = -C_m v_h - C_4 e^{-sT_d} v_e + C_6 F_h - C_2 e^{-sT_d} F_e \tag{9.15}$$

$$F_{cs} = -C_s v_e + C_1 e^{-sT_d} v_h + C_3 e^{-sT_d} F_h - C_5 F_e. \tag{9.16}$$

Taking into account

$$v_h = \frac{F_h + F_{cm}}{Z_m} \quad \Rightarrow \quad F_{cm} = -F_h + Z_m v_h \tag{9.17}$$

$$v_e = \frac{-F_e + F_{cs}}{Z_s} \quad \Rightarrow \quad F_{cs} = F_e + Z_s v_e \tag{9.18}$$

the dynamics of the teleoperation system can be written as

$$Z_{cm} v_h + C_4 e^{-sT_d} v_e = (1 + C_6) F_h - C_2 e^{-sT_d} F_e \tag{9.19}$$

$$Z_{cs} v_e - C_1 e^{-sT_d} v_h = C_3 e^{-sT_d} F_h - (1 + C_5) F_e, \tag{9.20}$$

where $Z_{cm} = Z_m + C_m$ and $Z_{cs} = Z_s + C_s$. In order to complete stability and performance analysis of a teleoperation system, we first need to obtain hybrid parameters of immittance mapping based on (8.5) from Eqs. (9.19) and (9.20).

Parameter h_{11} is defined as $h_{11} = \frac{F_h}{v_h}|_{F_e=0}$. From Eq. (9.20) we express

$$v_e|_{F_e=0} = \frac{1}{Z_{cs}} (C_3 e^{-sT_d} F_h + C_1 e^{-sT_d} v_h) \tag{9.21}$$

and by inserting the above expression into Eq. (9.19) we obtain

$$h_{11} = \frac{Z_{cm}Z_{cs} + C_1C_4e^{-2sT_d}}{(1+C_6)Z_{cs} - C_3C_4e^{-2sT_d}}. \tag{9.22}$$

Parameter h_{12} is defined as $h_{12} = \frac{F_h}{F_e}|_{v_h=0}$. From Eq. (9.20) we express

$$v_e|_{v_h=0} = \frac{1}{Z_{cs}}(C_3e^{-sT_d}F_h - (1-C_5)F_e) \tag{9.23}$$

and by inserting the above expression into Eq. (9.19) we obtain

$$h_{12} = \frac{C_2Z_{cs}e^{-sT_d} - C_4(1+C_5)e^{-sT_d}}{(1+C_6)Z_{cs} - C_3C_4e^{-2sT_d}}. \tag{9.24}$$

Parameter h_{21} is defined as $h_{21} = -\frac{v_e}{v_h}|_{F_e=0}$. From Eq. (9.19) we express

$$F_h|_{F_e=0} = \frac{1}{1+C_6}(Z_{cm}v_h + C_4e^{-sT_d}v_e) \tag{9.25}$$

and by inserting the above expression into Eq. (9.20) we obtain

$$h_{21} = -\frac{C_3Z_{cm}e^{-sT_d} + C_1(1+C_6)e^{-sT_d}}{(1+C_6)Z_{cs} - C_3C_4e^{-2sT_d}}. \tag{9.26}$$

Parameter h_{22} is defined as $h_{22} = -\frac{v_e}{F_e}|_{v_h=0}$. From Eq. (9.19) we express

$$F_h|_{v_{h=0}} = \frac{1}{1+C_6}(C_2e^{-sT_d}F_e + C_4e^{-sT_d}v_e) \tag{9.27}$$

and by inserting the above expression into Eq. (9.20) we obtain

$$h_{22} = \frac{(1+C_5)(1+C_6) - C_2C_3e^{-2sT_d}}{(1+C_6)Z_{cs} - C_3C_4e^{-2sT_d}}. \tag{9.28}$$

By inserting the parameters ($h_{ij}: i, j \in \{1, 2\}$) into Eq. (9.8) we obtain the impedance perceived by the operator

$$Z_{to} = \frac{(Z_{cm}Z_{cs} + C_1C_4e^{-2sT_d}) + [(1+C_5)Z_{cm} + C_1C_2e^{-2sT_d}]Z_e}{[(1+C_6)Z_{cs} - C_3C_4e^{-2sT_d}] + [(1+C_5)(1+C_6) - C_2C_3e^{-2sT_d}]Z_e}. \tag{9.29}$$

A simplified model of a four-channel architecture with transfer function (9.29) is shown in Fig. 9.4.

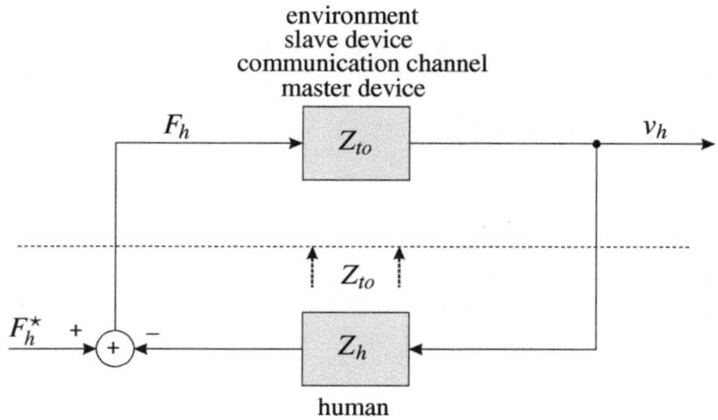

Fig. 9.4 A simplified model of a four-channel teleoperation system architecture

In conditions of negligible time delay T_d it is possible to design a control system that guarantees perfect transparency (conditions 9.9) of the teleoperation system using the following parameters:

$$\begin{cases} C_1 = Z_{cs} \\ C_2 = 1 + C_6, \quad C_2 \neq 0 \\ C_3 = 1 + C_5, \quad C_3 \neq 0 \\ C_4 = -Z_{cm} \end{cases} \tag{9.30}$$

In conditions of perfect transparency the equality $Z_{to} = Z_e$ holds. A physical interpretation of the choice of transfer functions (9.30) shows that a perfect transparency can be obtained if the dynamics of master and slave devices are perfectly compensated using models Z_{cm} and Z_{cs}, and forces that are the result of feedforward compensation have to match forces applied by the operator on the master device and the environment on the slave device. In such a case the master and slave devices are virtually removed from the system, resulting in direct virtual coupling between the operator and the environment.

The choice of controllers (9.30) guarantees perfect transparency of a teleoperation system. However, the most important characteristic of a teleoperation system is its stability. Even though the system is stable for values $C_2,\ C_3 > 0$, since $h_{11} = h_{22} = 0$ and $h_{12} = -h_{21} = 1$, the control is not robust as the system lies in the region of marginal stability with $\eta_{\mathscr{H}} = 1$.

In the absence of time delays, analysis of stability and efficiency is relatively simple and perfect transparency can be achieved. In the case of longer time delays it is necessary to find a compromise between system stability and transparency.

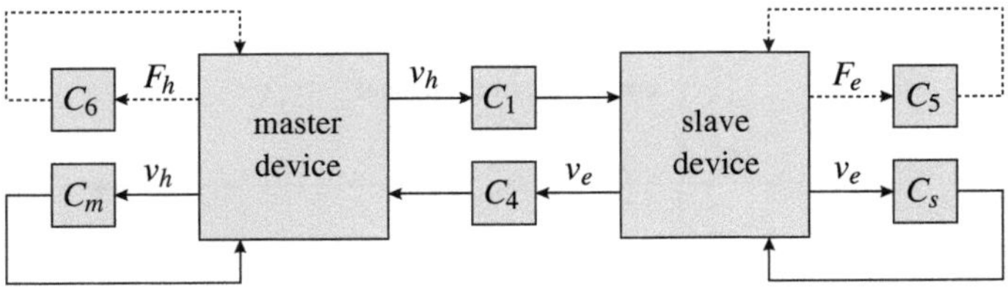

Fig. 9.5 Two-channel control architecture with $C_2 = C_3 = 0$. Control variables important for system stability are indicated with a *solid line*, while a *dashed line* is used to indicate complementary variables. Adapted from [5]

9.4 Two-Channel Control Architectures

Four-channel control architectures shown in Fig. 9.3 can be simplified if only one signal is led from a master to a slave device and vice versa—position or force [5]. Four different two-channel control architectures can be defined based on the quantities transmitted from the master to the slave device and quantities measured on the slave device (forces or velocities). Since two-channel control architectures require smaller number of sensors and are simpler than four-channel control architectures, their use prevails. Elimination of two signal pathways simplifies analytical analysis of two-channel architectures.

Stability and efficiency of a teleoperation system based on two-channels control architectures will be analyzed in the presence of time delays. Closed loop dynamics of slave device (9.20) can be rewritten as

$$\frac{1}{Z_{cs}} F_e = -v_e + \frac{1}{Z_{cs}} [-C_5 F_e + C_1 e^{-sT_d} v_h + C_3 e^{-sT_d} F_h]. \tag{9.31}$$

By comparing the above Eq. (9.31) with the dynamics of an admittance type slave (9.14), we obtain $Y_s = \frac{1}{Z_{cs}}$ and $v_{cs} = \frac{1}{Z_{cs}}(-C_5 F_e + C_1 e^{-sT_d} v_h + C_3 e^{-sT_d} F_h)$. Impedance Z_{cs} is usually high with the use of a robot as a slave device. $\frac{1}{Z_{cs}}$ represents a low admittance of the slave device, which can be controlled using a position controller C_s. The control architecture shown in Fig. 9.5 can be implemented if the force feedforward controllers transfer functions equal zero ($C_2 = C_3 = 0$). Hybrid parameters in Eqs. (9.22), (9.24), (9.26) and (9.28) can be simplified to

$$h_{11} = \frac{Z_{cm} Z_{cs} + C_1 C_4 e^{-2sT_d}}{(1 + C_6) Z_{cs}} \tag{9.32}$$

$$h_{12} = \frac{-C_4 (1 + C_5) e^{-sT_d}}{(1 + C_6) Z_{cs}} \tag{9.33}$$

$$h_{21} = -\frac{C_1 e^{-sT_d}}{Z_{cs}} \tag{9.34}$$

$$h_{22} = \frac{1 + C_5}{Z_{cs}}. \tag{9.35}$$

In order to simplify stability analysis, hybrid parameters can be replaced with impedance parameters

$$z_{11} := \frac{\Delta h}{h_{22}} = \frac{Z_{cm}}{1 + C_6} \tag{9.36}$$

$$z_{12} := \frac{h_{12}}{h_{22}} = \frac{-C_4 e^{-sT_d}}{1 + C_6} \tag{9.37}$$

$$z_{21} := \frac{-h_{12}}{h_{22}} = \frac{C_1 e^{-sT_d}}{1 + C_5} \tag{9.38}$$

$$z_{22} := \frac{1}{h_{22}} = \frac{Z_{cs}}{1 + C_5}. \tag{9.39}$$

Stability parameter (9.4) can then be computed as

$$\begin{aligned}
\eta_p(\omega) &= \eta_{p_1} + \eta_{p_2} \\
&= -\cos\left(\angle \frac{-C_1 C_4 e^{-j2\omega T_d}}{(1 + C_5)(1 + C_6)}\right) + 2\frac{\Re e\left\{\frac{Z_{cm}}{1+C_6}\right\} \Re e\left\{\frac{Z_{cs}}{1+C_5}\right\}}{\left|\frac{C_1 C_4 e^{-j2\omega T_d}}{(1+C_5)(1+C_6)}\right|} \\
&= -\frac{\Re e\left\{\frac{-C_1 C_4 e^{-j2\omega T_d}}{(1+C_5)(1+C_6)}\right\}}{\left|\frac{-C_1 C_4 e^{-j2\omega T_d}}{(1+C_5)(1+C_6)}\right|} + 2\frac{\Re e\left\{\frac{Z_{cm}}{1+C_6}\right\} \Re e\left\{\frac{Z_{cs}}{1+C_5}\right\}}{\left|\frac{C_1 C_4 e^{-j2\omega T_d}}{(1+C_5)(1+C_6)}\right|} \\
&= \operatorname{sgn}\left((1 + C_5)(1 + C_6)\right)\left[-\cos(\angle - C_1 C_4 e^{-j2\omega T_d}) + 2\frac{\Re e\{Z_{cm}\}\Re e\{Z_{cs}\}}{|C_1 C_4|}\right].
\end{aligned} \tag{9.40}$$

Since $|e^{-j2\omega T_d}| = 1$, η_{p_2} does not depend on the time delay. Thus, the effect of the delay remains limited to the parameter η_{p_1}. Since $\eta_{p_1} \in [-1, 1]$, it is possible to guarantee unconditional stability only by satisfying condition $n_{p_2} \geq 2$, meaning

$$\operatorname{sgn}((1 + C_5)(1 + C_6))|C_1 C_4| \leq \Re e\{Z_{cm}\}\Re e\{Z_{cs}\}. \tag{9.41}$$

The above condition shows that at least minimal physical damping is required in the master and slave devices. In order to guarantee stability when decreasing the damping in the master device it is necessary to increase damping in the slave device and vice versa. Higher damping or velocity feedback loop inside the master or slave devices and lower impedances C_1 and C_4 increase the system stability margin. Since product $\Re e\{Z_{cm}\}\Re e\{Z_{cs}\}$ is not frequency dependent, while product $|C_1 C_4|$ is, the above condition can be satisfied only in a certain frequency range. The effect of local position controllers C_m and C_s on system stability is more important than the effect

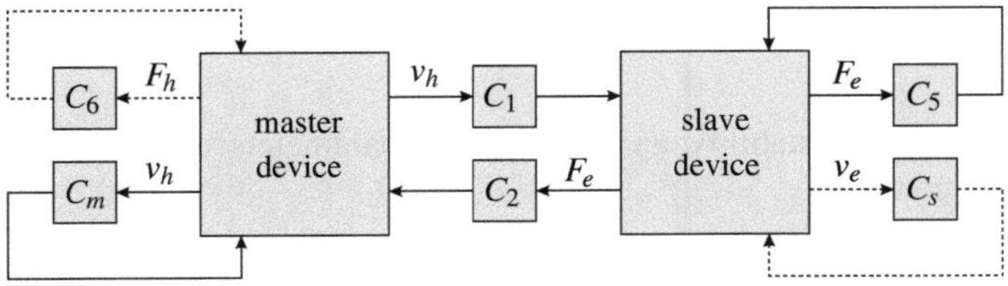

Fig. 9.6 Two-channel control architecture with $C_3 = C_4 = 0$. Control variables important for system stability are indicated with a *solid line*, while a *dashed line* is used to indicate complementary variables. Adapted from [5]

of local force controllers C_5 and C_6. Unconditional stability of uncoupled master (z_{11}) and slave (z_{22}) devices is satisfied, since impedances Z_{cm} and Z_{cs} are passive.

Z-width of the teleoperation system can be determined based on (9.10) as

$$\begin{aligned} Z_{to_{min}} &= \frac{Z_{cm}}{1+C_6} - \frac{-C_1C_4e^{-2sT_d}}{(1+C_6)Z_{cs}} \\ Z_{to_{max}} &= \frac{Z_{cm}}{1+C_6}. \end{aligned} \tag{9.42}$$

Lower impedances C_1 and C_4 increase stability margin (9.41) and reduce Z-width of a teleoperation system. An increase of local force controller gain C_6 decreases $Z_{to_{min}}$ as well as Z-width. A decrease of the local position controller gains C_m ($Z_{cm} = Z_m + C_m$) and C_s ($Z_{cs} = Z_s + C_s$) increases efficiency during movement in free space and increases Z-width of a teleoperation system at a cost of reduced stability.

A different two-channel control architecture can be obtained by setting $C_3 = C_4 = 0$ as shown in Fig. 9.6.

The master device force feedforward controller and slave device velocity feedforward controller do not appear in the control architecture. Hybrid parameters from Eqs. (9.22), (9.24), (9.26) and (9.28) can be rewritten as

$$h_{11} = \frac{Z_{cm}}{1+C_6} \tag{9.43}$$

$$h_{12} = \frac{C_2e^{-sT_d}}{1+C_6} \tag{9.44}$$

$$h_{21} = \frac{-C_1e^{-sT_d}}{Z_{cs}} \tag{9.45}$$

$$h_{22} = \frac{1+C_5}{Z_{cs}}. \tag{9.46}$$

Stability parameter (9.4) can be computed by replacing general immittance parameters with hybrid parameters, leading to

$$
\begin{aligned}
\eta_p(\omega) &= \eta_{p1} + \eta_{p2} \\
&= \cos\left(\angle \frac{C_1 C_2 e^{-j2\omega T_d}}{(1+C_6)Z_{cs}}\right) + 2\frac{\Re e\left\{\frac{Z_{cm}}{1+C_6}\right\}\Re e\left\{\frac{1+C_5}{Z_{cs}}\right\}}{\left|\frac{-C_1 C_2 e^{-j2\omega T_d}}{(1+C_6)Z_{cs}}\right|} \\
&= \mathrm{sgn}(1+C_6)\left[\cos\left(\angle \frac{C_1 C_2 e^{-j2\omega T_d}}{Z_{cs}}\right) + 2\frac{(1+C_5)\Re e\{Z_{cm}\}}{|C_1 C_2|}\cos\left(\angle \frac{1}{Z_{cs}}\right)\right],
\end{aligned}
\tag{9.47}
$$

where $\cos(\angle \frac{1}{Z}) = \cos(\angle Z)$. As for the control architecture in Fig. 9.5, the unconditional stability criterion is satisfied only in a limited frequency range, where $\eta_{p2} \geq 2$, meaning

$$
(1+C_5)\Re e\{Z_{cm}\} \geq \mathrm{sgn}(1+C_6)|C_1 C_2|. \tag{9.48}
$$

In order to increase stability margin of a teleoperation system an increased damping of master device and increased gain of the slave device local force controller (higher master device impedance and lower slave device impedance) are required, while at the same time the parameters of the feedforward controllers C_1 and C_2 need to be reduced. From condition (9.48) we can conclude that at least minimal physical damping is required in the master device. The master device local force controller C_6 and the slave device local position controller C_s are not significant for stability analysis. Finally, we can observe that the master device physical damping is more important for system stability than the slave device physical damping.

Using the hybrid parameters (9.43)–(9.46) we can define the Z-width of the teleoperation system based on (9.10) as

$$
\begin{aligned}
Z_{to_{min}} &= \frac{Z_{cm}}{1+C_6} \\
Z_{to_{max}} &= \frac{Z_{cm}}{1+C_6} + \frac{C_1 C_2 e^{-2sT_d}}{(1+C_5)(1+C_6)}.
\end{aligned}
\tag{9.49}
$$

Similar to the control architecture shown in Fig. 9.5, an increase of feedforward controller parameters (C_1 and C_2) improves system transparency and Z-width. An increase of the local force control loop gains C_5 and C_6 decreases $Z_{to_{min}}$ and Z-width. A decrease of the master device position controller parameters C_m improves system performance in free space at the cost of decreasing system stability.

Considering stability parameters (9.40), (9.47) and Z-width (9.42), (9.49), general guidelines for design of two-channel control architectures can be defined. System stability can be increased by decreasing the gains of feedforward controller transfer functions, which also results in a decrease of the dynamic range of the device and/or higher minimal transmitted impedance. It is therefore necessary to find a compromise between stability and efficiency of the system. Regardless of the choice of feedforward controller transfer functions, when a measured force (position) signal

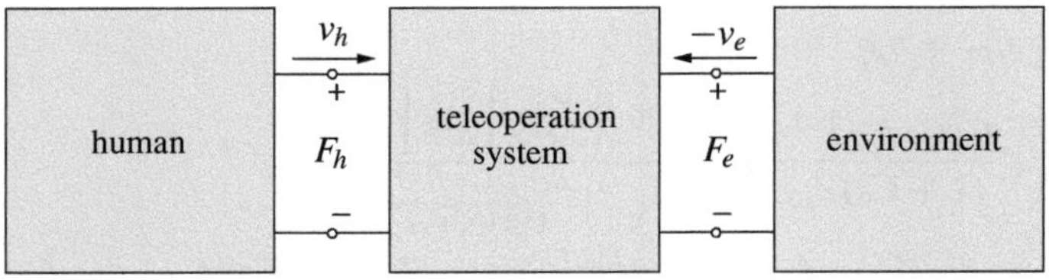

Fig. 9.7 Two-port network model of a teleoperation system. Adapted from [6]

is transmitted via a communication channel, the local position (force) control loop on the side of the transmitter is more important for system stability than the local position (force) control loop. An increase of values of the local force control loop gains consequently reduces $Z_{to_{min}}$ as well as Z-width. An increase of values of the local position control loop gains increases $Z_{to_{min}}$.

9.5 Passivity of a Teleoperation System

In Sect. 8.7 we analyzed passivity of a haptic interface in time domain and we found that system passivity can be guaranteed by implementing passivity observer and passivity controller. Results of the analysis will now be applied also to teleoperation systems [6].

Figure 9.7 shows a model of a teleoperation system, where v_h and v_e define velocities of the master and slave devices at the points of interaction, F_h is the force applied by the operator on the master device and F_e is the force applied by the slave device on the environment.

Stability of the teleoperation system will be analyzed from the passivity perspective. If individual elements constituting the teleoperation system are passive, then also the complete system will behave passively. In general, we can assume that the environment is passive and that the human operator behaves passively as well. Thus, if a two-port network representing the teleoperation system is passive, the system in Fig. 9.7 will be passive as well.

Definition 9.1 An M-port network (Fig. 9.8c) with the initial stored energy $E(0)$ is passive if and only if

$$\int_0^t (F_1(\tau)v_1(\tau) + \cdots + F_M(\tau)v_M(\tau))\, d\tau + E(0) \geq 0, \quad \forall t \geq 0 \tag{9.50}$$

for admissible forces $(F_1, \ldots, F_M)$ and velocities $(v_1, \ldots, v_M)$. Signs of velocities and forces are defined such that their product is positive if energy enters the system.

The passivity observer and controller were introduced in Sect. 8.7. If the initial stored energy of a one-port network equals zero and it is possible to measure quantities

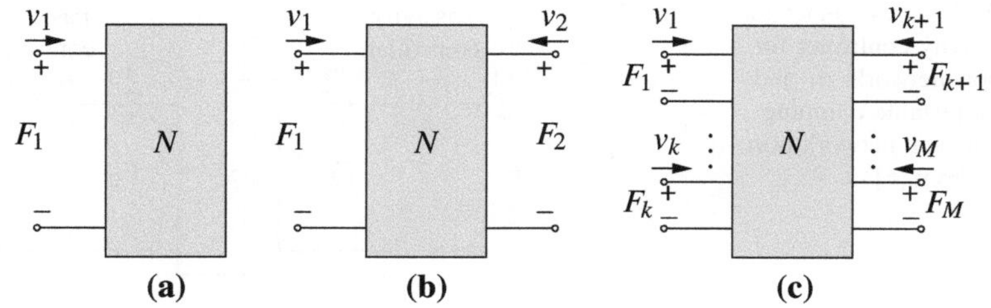

Fig. 9.8 **a** One-port network **b** two-port network and **c** multi-port network. Adapted from [6]

F and v that define the energy flow into the system with a sampling frequency that is significantly higher than the system dynamics, so that the changes of force and velocity within a sampling interval are small, it is possible to equip one or more elements of the teleoperation system with a passivity observer that measures the flow of energy into a one-port network shown in Fig. 9.8a as

$$E_{obsv}(n) = T\sum_{k=0}^{n} F_1(k)v_1(k), \tag{9.51}$$

where T is the sampling time. If $E_{obsv}(n) \geq 0$ for all n, then the system consumes energy. In the case that $E_{obsv}(n) < 0$, the system generates energy and the quantity of generated energy equals $-E_{obsv}(n)$.

Depending on the operating conditions and the dynamics of the one-port network, the output of the passivity observer can be positive or negative at a specific time instance. If it is negative at any time, the one-port network contributes to unstable behavior of the system. At the same time the amount of produced energy is known. Therefore, it is possible to design a time dependent element that will consume the excessive energy—passivity controller.

The concept of passivity observer and controller will be extended to two-port networks with the goal of guaranteeing passivity of a teleoperation system. Similarly as for a one-port network, we can define a passivity observer for a two-port network shown in Fig. 9.8b as

$$E_{obsv}(n) = T\sum_{k=0}^{n}(F_1(k)v_1(k) + F_2(k)v_2(k)). \tag{9.52}$$

The passivity controller needs to be designed for each input port and it is necessary to activate the passivity controller on the input port that is active.

If a single passivity controller is used for the entire two-port network and connected to the input port (F_1, v_1), it might happen that $E_{obsv}(n) < 0$, even though $v_1 = 0$ or $F_1 = 0$. In this case the energy generated at the input port (F_2, v_2) cannot be consumed, as the only passivity controller attached to the input port (F_1, v_1)

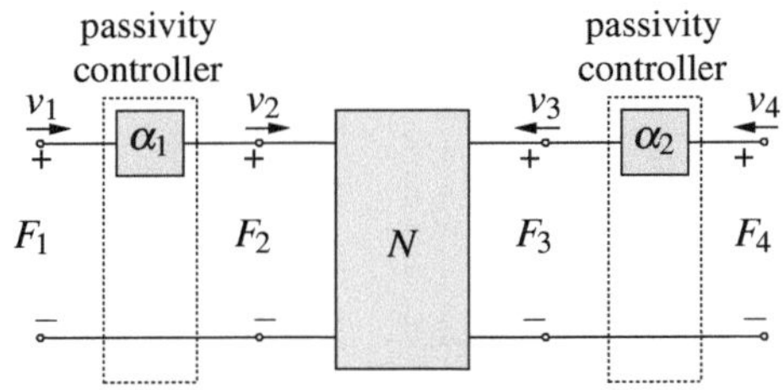

Fig. 9.9 A series connection of a passivity controller for a two-port network. α_1 and α_2 are adjustable damping values on each network port. Adapted from [6]

cannot be activated with zero input signals. Therefore, the passivity controller attached to the active port needs to be activated. Passivity or activity of a network port can simply be verified by observing the port input variables

$$\text{input port condition} = \begin{cases} \text{active}, & \Leftarrow \quad Fv < 0 \\ \text{passive}, & \Leftarrow \quad Fv \geq 0. \end{cases} \tag{9.53}$$

For an impedance type of two-port network the passivity controllers α_1 and α_2 (Fig. 9.9) can be computed in real time by assuming that $v_1(n) = v_2(n)$ and $v_3(n) = v_4(n)$ are system inputs, while $F_2(n)$ and $F_3(n)$ are system outputs. Then the passivity observer equals

$$E_{obsv}(n) = T \sum_{k=0}^{n-1} (F_1(k)v_1(k) + F_4(k)v_4(k)) + F_2(n)v_2(n) + F_3(n)v_3(n). \tag{9.54}$$

Different cases need to be considered when computing α_1 and α_2. If $E_{obsv}(n) \geq 0$, energy does not flow out of the system and there is no need for activation of the passivity controller, leading to $\alpha_1 = \alpha_2 = 0$.

If $E_{obsv}(n) < 0$, $F_2(n)v_2(n) < 0$ and $F_3(n)v_3(n) \geq 0$, energy flows out of the system through the port (F_2, v_2). Therefore, it is necessary to activate the passivity controller $\alpha_1 = \frac{-E_{obsv}(N)}{Tv_2(n)^2}$, while $\alpha_2 = 0$.

If $E_{obsv}(n) < 0$, $F_2(n)v_2(n) \geq 0$ and $F_3(n)v_3(n) < 0$, energy flows out of the system through the port (F_3, v_3). Therefore, it is necessary to activate the passivity controller $\alpha_2 = \frac{-E_{obsv}(N)}{Tv_3(n)^2}$, while $\alpha_1 = 0$.

If $E_{obsv}(n) < 0$, $F_2(n)v_2(n) < 0$ and $F_3(n)v_3(n) < 0$, energy flows out of the system through both ports. Therefore, it is necessary to activate both passivity controllers that together need to consume the superfluous energy $-E_{obsv}(n)$. For an impedance causal structure the damping needs to be distributed between the two input ports, such that $T(\alpha_1(n)v_1(n)^2 + \alpha_2(n)v_2(n)^2) = -E_{obsv}(n)$.

System outputs can then be computed as

$$\begin{aligned} F_1(n) &= F_2(n) + \alpha_1(n)v_2(n) \\ F_4(n) &= F_3(n) + \alpha_2(n)v_3(n). \end{aligned} \tag{9.55}$$

With the implementation of the above algorithm, passive system behavior can be easily demonstrated:

$$
\begin{aligned}
&T\sum_{k=0}^{n}(F_1(k)v_1(k) + F_4(k)v_4(k)) \\
&= T\sum_{k=0}^{n}F_2(k)v_2(k) + T\sum_{k=0}^{n}F_3(k)v_3(k) + T\sum_{k=0}^{n}\alpha_1(k)v_2(k)^2 + T\sum_{k=0}^{n}\alpha_2(k)v_3(k)^2 \\
&= T\sum_{k=0}^{n}F_2(k)v_2(k) + T\sum_{k=0}^{n}F_3(k)v_3(k) + T\sum_{k=0}^{n-1}\alpha_1(k)v_2(k)^2 + T\sum_{k=0}^{n-1}\alpha_2(k)v_3(k)^2 \\
&\quad + T\alpha_1(n)v_2(n)^2 + T\alpha_2(n)v_3(n)^2 \\
&= E_{obsv}(n) + T\alpha_1(n)v_2(n)^2 + T\alpha_2(n)v_3(n)^2.
\end{aligned}
\tag{9.56}
$$

By taking into account passivity controllers α_1 and α_2 it is possible to guarantee passivity of a teleoperation system under all operating conditions:

$$
\sum_{k=0}^{n}(F_1(k)v_1(k) + F_4(k)v_4(k)) \geq 0, \quad \forall n. \tag{9.57}
$$

For a teleoperation system with an admittance structure, passivity can be controlled with parallel connection of a passivity controller.

In order to determine an appropriate location for passivity observer and controller, it is necessary to analyze accessibility of pairs of signals (F, v) on individual ports. In addition to being accessible in real time, the signal pair must allow modification of signal values using the passivity controller. In general it is possible to measure forces (F_h and F_e) and velocities (v_h and v_e) on both input ports of a two-port network. However, these signals cannot be modified in real time, since these are the result of a physical interaction between the human/environment and the teleoperation system.

In order to implement a passivity controller it is necessary to virtually remove certain passive elements from the system and find a reachable pair of variables (F, v) without loss of passivity of the entire system. In physical terms, energy flows into the mechanical system from the point of actuation and can be computed from a pair of variables that define the energy flow from the actuators, as for example force at the actuator output and the velocity in the point of actuation. The entire teleoperation system can be split into three parts: (1) the mechanism of the master device, (2) controller of master and slave devices with the addition of the communication channel and (3) mechanism of the slave device. Figure 9.10 shows the complete model of a teleoperation system with individual energy flows. The bilateral controller exchanges energy with the mechanisms of master and slave devices and these energy flows can be measured across the pairs of variables (F_m, v_m) and (F_s, v_s). F_m and F_s are the two forces that actuate mechanisms of master and slave devices and v_m and v_s are velocities at the actuation points. Mechanisms of master and slave devices can be

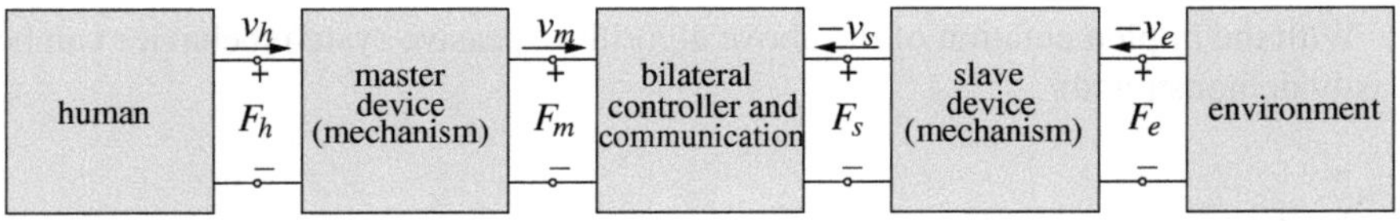

Fig. 9.10 A model of a teleoperation system in interaction with a human and the environment. Adapted from [6]

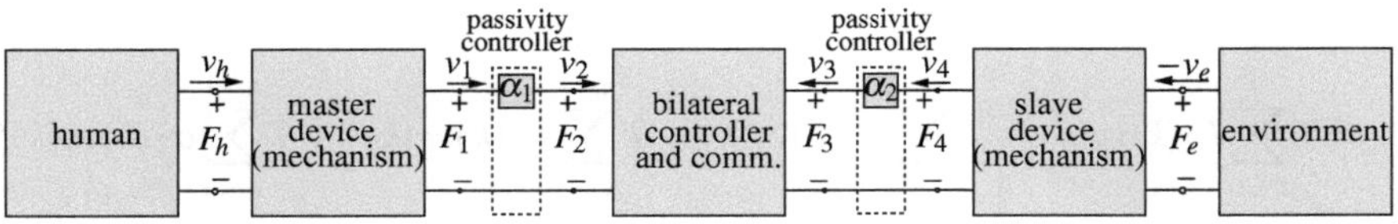

Fig. 9.11 A model of a teleoperation system with a passivity controller in interaction with a human and the environment

removed from the analysis due to their passive nature, since they do not contribute to the active behavior of a teleoperation system. The passivity observer can be defined as

$$E_{obsv}(n) = T \sum_{k=0}^{n} (F_m(k)v_m(k) + F_s(k)v_s(k)) \tag{9.58}$$

and placed at the two-port network representing the bilateral controller.

Inputs to the bilateral controller are thus defined as velocities of master (v_m) and slave (v_s) devices while outputs of the bilateral controller are control variables for master (F_m) and slave (F_s) devices. Thus, a passivity controller can be implemented as a series connection as shown in Fig. 9.11.

References

1. Lawrence, D.A.: Stability and transparency in bilateral teleoperation. IEEE Trans. Robot. Autom. **9**, 624–637 (1993)
2. Yokokohji, Y., Yoshikawa, T.: Bilateral control of master-slave manipulators for ideal kinesthetic coupling—formulation and experiment. IEEE Trans. Robot. Autom. **10**, 605–620 (1994)
3. Cavusoglu, M.C., Feygin, D.: Kinematics and dynamics of Phantom model 1.5 haptic interface. Technical report, Department of Electrical Engineering and Computer Sciences, University of California, Berkeley (2001)
4. Hannaford, B.: A design framework for teleoperators with kinesthetic feedback. IEEE Trans. Robot. Autom. **5**, 426–434 (1989)
5. Hashtrudi-Zaad, K., Salcudean, S.E.: Analysis of control architectures for teleoperation systems with impedance/admittance master and slave manipulators. Int. J. Robot. Res. **20**, 419–445 (2001)
6. Ryu, J.H., Kwon, D., Hannaford, B.: Stable teleoperation with time-domain passivity control. IEEE Trans. Robot. Autom. **20**, 365–373 (2004)

Chapter 10
Virtual Fixtures

Virtual fixtures are software generated movement-constraints used in teleoperation and cooperative robotic systems. As the name implies, virtual fixtures are not mechanical constraints of the robotic system or the environment with which the robotic system is in contact, but they are instead part of the control algorithms and are overlaid on the physical environment in the workspace of the robotic system. In teleoperation and cooperative systems the robotic device responds to movement commands of the human operator and the main purpose of the virtual fixtures is to guide or to confine movements commanded by the human operator. Virtual fixtures are used in applications where a task requires both precision and accuracy of the robotic system as well as the decision making power and intelligence of the human operator [1].

A common real world metaphor for virtual fixtures is use of a ruler for drawing a straight line [2]. Drawing a straight line requires good hand-eye coordination, where visual input from eyes and proprioceptive input from hands is combined to guide the neuromuscular system. This requires a certain level of mental effort by the person. However, if a ruler is used to guide the hand, drawing a line requires much less mental effort. The task is executed much faster and more accurately. In the same way as a ruler is put on top of a piece of paper, the virtual fixture is overlaid on the physical environment to enhance the operator's performance. Since the virtual fixtures are software generated, both geometrical and interactive properties of the virtual fixture can be chosen according to the specific requirements of the task [3].

One benefit of virtual fixtures is that their shape can be adapted to the specific requirements of the task, while another benefit is that there is no actual physical entity involved in the interaction with the environment. Rosenberg [2] defined the virtual fixture as *abstract sensory information overlaid over the reflected sensory feedback from an environment with which the manipulator of the teleoperation or collaborative robotic system is in contact*. Although these virtual constraints are software-generated, the operator feels them as mechanical or haptic constraints due to the constrained motion of the robotic manipulator.

M. Mihelj and J. Podobnik, *Haptics for Virtual Reality and Teleoperation*,
Intelligent Systems, Control and Automation: Science and Engineering 64,
DOI: 10.1007/978-94-007-5718-9_10,

Virtual fixtures provide a balance between robotic systems that are under full direct control of the human operator and robotic systems that are fully autonomous. They are especially suited for applications where solely unaided human accuracy and performance is inadequate, but on the other hand, a human intelligence is required due to inadequate capabilities of artificial intelligence of autonomous systems [4].

10.1 Types of Virtual Fixtures

Virtual fixtures can be categorized into two basic categories: guidance virtual fixtures and forbidden-region virtual fixtures [1]. Guidance virtual fixtures provide assistance to the operator in moving a robot manipulator along desired paths or surfaces in the workspace of the manipulator, while forbidden-region virtual fixtures prevent motion into the forbidden region. Figure 10.1 shows basic properties of virtual fixtures.

- Guidance virtual fixtures—guide and assist the movement along the desired path, direction or surface in the allowed workspace of the robotic system. Guidance virtual fixtures aim to increase the accuracy and performance of the human operator.
- Forbidden-region virtual fixtures—prevent motion into the forbidden region of workspace and are meant as safety movement constraints. Forbidden-region virtual fixtures aim to increase the safety of operation of a robotic system and prevent undesired deviations into a restricted region of workspace, where the operation of the robotic system could lead to an unsafe interaction between the robotic system and the environment.

The majority of applications of virtual fixtures are implemented in two types of human-robot collaborative robotic systems:

- Teleoperation robotic systems (see Chap. 9 for details). In teleoperation systems the slave device is in contact with the remote environment and responds to movements

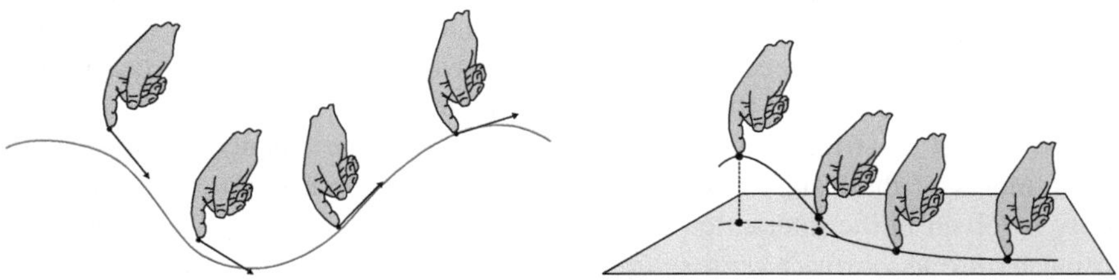

(a) Guidance virtual fixture **(b)** Forbidden-region virtual fixture

Fig. 10.1 Illustration of the principle of virtual fixtures. **a** The principle of guidance virtual fixtures. The tool tip is guided along the desired path. **b** A principle of forbidden-region virtual fixtures. The tool tip is free to move in the safe region of the workspace, while the movement directed into the unsafe region of the workspace is constrained

of the master device. The master device is operated by the human operator. Teleoperation systems therefore typically consist of two robotic manipulators. Virtual fixtures can be applied both on slave and master devices of the teleoperation robotic system.

- Cooperative robotic systems. In cooperative robotic systems the operator directly guides the robotic manipulator, which is in contact with and manipulates the environment. Most notable examples of cooperative robotic systems are cobots [5] and human-machine cooperative systems [4].

In both types of human-robot collaborative systems the human operator is "in-the-loop" with the robotic manipulator, directly or remotely operating with the manipulator device.

10.1.1 Cobots

The motivation for development of cobots was a need for computer-controlled mechanical assistance to human workers [4]. This resulted in the concept of a human operator and a collaborative robot manipulator, in short cobot, working together, physically linked through a manipulated object. Cobots are therefore intended for a collaborative work with human operators. The manipulated object is in contact (grasped) by both the robotic manipulator and the human operator. The robotic manipulator of the cobots and a human operator thus share the workspace.

Cobots are inherently passive and move only when the movement is exerted by the human operator, which is responsible for generating sufficient force and moments needed for execution of movements. Passivity of the cobots ensures safety of the interaction. Both the guidance virtual fixtures and forbidden-region virtual fixtures can be used with cobots. The task of the cobot is to shape and guide the movement according to a predefined path or to restrict it to a predefined region of the shared human/cobot workspace.

In case of active devices the torque produced by the motors in the joints of the robotic manipulator is mapped through the kinematic structure of the manipulator to the contact point to produce the force, which constrains and guides the movement. However, cobots, which are passive, do not use motors to produce forces, which would resist the motion directed by the human operator. The resistance is instead controlled with steerable transmissions and the reaction force is transferred to the manipulator's base fixed on the ground. The simplest example of a cobot is a three-wheeled device. If two wheels are fixed and the third wheel is steered along a curved path in the $x - y$ plane, the movement of the human operator will be constrained to that path.

10.1.2 Human-Machine Cooperative Systems

In contrast to the cobots, the human-machine cooperative systems utilize a robotic manipulator that can actively generate motion. Furthermore, in contrast to the teleoperation robotic systems the operator is in direct control of the robotic manipulator by either holding the tool mounted to the robotic manipulator or by holding the robotic manipulator itself. Virtual fixtures act on the tool tip position to guide or constrain the motion of the tool. Consequently, the operator, who is directly in contact with the manipulator, perceives the virtual fixtures as motion constraints. A typical robotic manipulator in a human-machine cooperative system is an admittance type device equipped with a force sensor for measuring the forces applied by the human operator and the environment. The robotic manipulator generates movement proportional to the measured forces by employing different control modes embedded into the controller of the robotic system. Admittance controlled cooperative systems are especially suited for tasks that require slow and precise movements at the limits of the human neuromotor abilities [1, 6].

Cooperative systems have been developed to aid the human operator to perform high-precision or microscale manipulation with reduced cognitive strain on the operator. Examples of such manipulation are microscale assembly tasks and surgical procedures, which require either microscale manipulation or precise microscale movements along anatomical features [7]. Collaboration improves the performance of the operator in both speed and accuracy when executing the task. Cooperative robotic manipulators compensate for the limitations of the human operator by suppressing the undesirable effects of muscle tremor, motion inaccuracy and positional drift. By employing virtual fixtures, the cooperative systems have the ability to guide and constrain movements commanded by the operator to improve accuracy and safety [8, 9].

The most common use of virtual fixtures in cooperative systems is in surgical robotic systems. The motivation for development of cooperative surgical robotic systems was the same as for development of teleoperation surgical robotic systems: increased accuracy, decreased time required to finish the task, decreased number of errors or potentially unsafe situations, all of which augment operator capabilities. Surgical robotic systems are especially suitable for use in minimally invasive surgery. Advantages compared to the teleoperation robotic systems are: simpler mechanical design of the robotic manipulators, simpler control architecture, easier adaptation of the surgical workstation and an ability to retain a level of proprioceptive feedback through direct control of the robotic manipulator. It is often emphasized that in precise and small scale manipulation, where the tactile information is inadequate due to hardly perceivable forces of the interaction between the tool and the environment, the force-feedback from the manipulator augments the manipulation capabilities of the operator and makes the manipulation more intuitive [1]. Advantages of the teleoperation systems compared to cooperative systems is the ability to scale forces and movements, which is not possible in cooperative systems and that only one robotic manipulator is used in cooperative systems compared to teleoperation systems where more robotic manipulators are used for a surgical procedure.

10.2 Guidance Virtual Fixtures

Guidance virtual fixtures are designed to help the operator to move the robotic manipulator along the preferred direction in a workspace of the robotic manipulator. A preferred direction can be: a direction toward a point, direction along a path, or a direction of movement along the surface of the virtual fixture. The aim is to augment the performance of the operator, such that the task is executed more accurately and faster at lower mental effort of the operator. Guidance virtual fixtures are characterized by geometrical properties and a control law.

The geometry of the virtual fixture defines the preferred direction of the movement in a task space. The movement is decomposed into two components: a preferred direction and a non-preferred direction that is perpendicular to the preferred direction. Although the geometry of the virtual fixture is task dependent, it can be represented by two basic geometrical primitives:

- A point in a task space (Fig. 10.2a): the virtual fixture is geometrically defined with the desired point $\mathbf{p}_d$ to which the robot manipulator should move. In a task in which a robot manipulator should move toward the desired point $\mathbf{p}_d$, the preferred direction is a direction from the current position of the robotic manipulator to the desired point $\mathbf{p}_d$.
- A direction along the path in a task space (Fig. 10.2b): the virtual fixture is geometrically defined by the direction of a unit vector $\mathbf{t}$. In a task in which the manipulator should move along the path, the preferred direction is given by a vector $\mathbf{t}$, which is the tangent vector at the closest point on the curve.

In a more complex task such as moving on a curved path, the preferred direction is defined by combining both geometrical primitives. A closest point on a curved path of the end-point of the manipulator is calculated by projecting the end-point position of the manipulator on the curve. A direction of a curved path is defined by calculating the tangent of a curve at the closest point. If a task requires the manipulator to move to a curve, the preferred direction would be a direction from an end-point of the manipulator to the closest point on the curve. If a task requires the robotic manipulator to move parallel with the curve, a preferred direction is then defined by the tangent at a closest point on the curve. However, the task typically requires a manipulator to move along the curved path closest as possible to the curve. The preferred direction is then the sum of two vectors: the first vector pointing to the closest point on the curve and the second vector pointing in the direction of a tangent at a closest point on the curve. The next section will cover how to calculate the closest point on a curved path of a guidance virtual fixture.

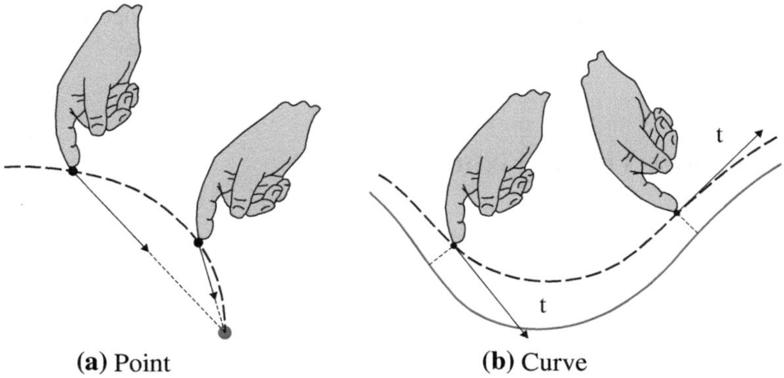

Fig. 10.2 The figure shows two basic geometrical primitives: a point and a path. **a** The movement toward a point. In any point of the workspace the movements of the tool tip are directed toward the desired point. **b** The movement along the path. At any point of the workspace the movements of the tool tip are directed to either follow the path or to take the direction of the tangent at the closest point on the *curve*

10.2.1 Tangent and Closest Point on the Curve

A curved path is defined by a differentiable parametric curve $\mathbf{r}(s)$ of a parameter s in three-dimensional task space. The curve is defined on the interval $s \subset [s_s, s_e]$ and singular points of the curve $\mathbf{r}(s)$ must not lie on the interval $u \subset [s_s, s_e]$. Singular points are points where a tangent at the curve becomes a zero vector.

The end-point position of the manipulator is given by $\mathbf{p}_e = [x_e, y_e, z_e]$. The end-point position can be written as a sum of vectors (Fig. 10.3)

$$\mathbf{p}_e = \mathbf{r}(s_0) + \Delta\mathbf{r} = \mathbf{r}(s_0 + \Delta s) + \mathbf{e}, \tag{10.1}$$

where s_0 is a known parameter value, and Δs is an unknown increment of parameter s, which is assumed to be small.

Vector $\mathbf{r}(s_0 + \Delta s)$ can be therefore written as

$$\mathbf{r}(s_0 + \Delta s) = \mathbf{r}(s_0) + \Delta\mathbf{r} - \mathbf{e}. \tag{10.2}$$

Using a Taylor series approximation, vector $\mathbf{r}(s_0 + \Delta s)$ can be written as

$$\mathbf{r}(s_0 + \Delta s) = \mathbf{r}(s_0) + \frac{\Delta s}{1!}\mathbf{r}'(s_0) + \frac{\Delta s^2}{2!}\mathbf{r}''(s_0) + \cdots . \tag{10.3}$$

Substituting the left side of Eq. (10.2) with a Taylor series approximation yields

$$\mathbf{r}(s_0) + \frac{\Delta s}{1!}\mathbf{r}'(s_0) + \frac{\Delta s^2}{2!}\mathbf{r}''(s_0) + \cdots = \mathbf{r}(s_0) + \Delta\mathbf{r} - \mathbf{e}. \tag{10.4}$$

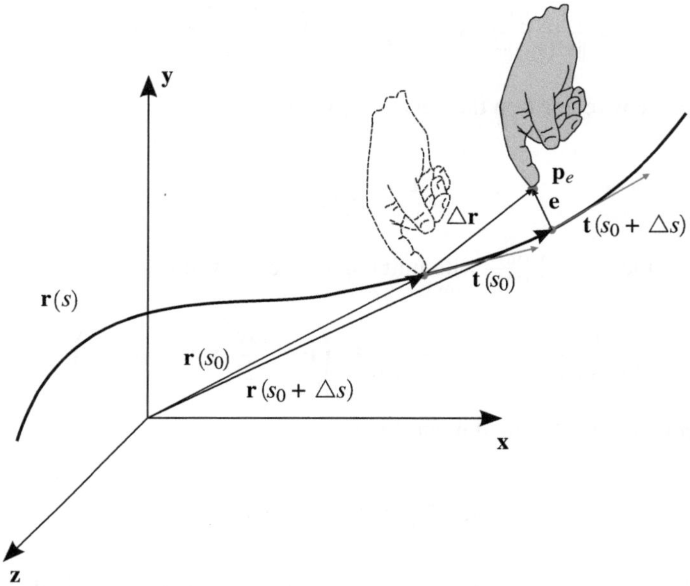

Fig. 10.3 Estimation of a tangent to a *curve* and the closest point on a *curve*

Subtracting the $\mathbf{r}(s_0)$ on both sides of the equation gives

$$\frac{\Delta s}{1!}\mathbf{r}'(s_0) + \frac{\Delta s^2}{2!}\mathbf{r}''(s_0) + \cdots = \Delta\mathbf{r} - \mathbf{e}. \tag{10.5}$$

Rearranging the expression (10.5) such that the higher order terms of Taylor series approximation are grouped together with error $\mathbf{e}$ on the right side of expression yields

$$\Delta s\mathbf{r}'(s_0) = \Delta\mathbf{r} - \left(\mathbf{e} + \frac{\Delta s^2}{2!}\mathbf{r}''(s_0) + \cdots\right). \tag{10.6}$$

Then both sides of the equation are multiplied by a vector $\mathbf{r}'(s_0)$ giving

$$\Delta s\mathbf{r}'(s_0)\cdot\mathbf{r}'(s_0) = \mathbf{r}'(s_0)\cdot\Delta\mathbf{r} - \mathbf{r}'(s_0)\cdot\left(\mathbf{e} + \frac{\Delta s^2}{2!}\mathbf{r}''(s_0) + \cdots\right) \tag{10.7}$$

$$\Delta s\|\mathbf{r}'(s_0)\|^2 = \mathbf{r}'(s_0)\cdot\Delta\mathbf{r} - \mathbf{r}'(s_0)\cdot\left(\mathbf{e} + \frac{\Delta s^2}{2!}\mathbf{r}''(s_0) + \cdots\right). \tag{10.8}$$

The incremental value Δs of a parameter s can then be expressed as

$$\Delta s = \frac{\mathbf{r}'(s_0)}{\|\mathbf{r}'(s_0)\|^2}\Delta\mathbf{r} + \frac{\mathbf{r}'(s_0)}{\|\mathbf{r}'(s_0)\|^2}\left(\mathbf{e} + \frac{\Delta s^2}{2!}\mathbf{r}''(s_0) + \cdots\right). \tag{10.9}$$

The unit tangent vector $\mathbf{t}$ on the curve in point $\mathbf{r}(s_0)$ is

$$\mathbf{t} = \frac{\mathbf{r}'(s_0)}{\|\mathbf{r}'(s_0)\|}. \tag{10.10}$$

Substituting the term $\frac{\mathbf{r}'(s_0)}{\|\mathbf{r}'(s_0)\|}$ with unit tangent vector $\mathbf{t}$ in (10.9) gives

$$\Delta s = \frac{1}{\|\mathbf{r}'(s_0)\|}\mathbf{t}\cdot\Delta\mathbf{r} + \frac{1}{\|\mathbf{r}'(s_0)\|}\mathbf{t}\cdot\left(\mathbf{e} + \frac{\Delta s^2}{2!}\mathbf{r}''(s_0) + \cdots\right). \tag{10.11}$$

The full equation for the increment Δs is

$$\Delta s = \frac{1}{\|\mathbf{r}'(s_0)\|}\mathbf{t}\cdot\Delta\mathbf{r} + \Delta s_{e,p} + \Delta s_{e,d}, \tag{10.12}$$

where $\Delta s_{e,p} = \frac{1}{\|\mathbf{r}'(s_0)\|}\mathbf{t}\cdot\mathbf{e}$ is an error term due to a non-perpendicularity of the tangent $\mathbf{t}$ and vector $\mathbf{e}$ and $\Delta s_{e,d} = \frac{1}{\|\mathbf{r}'(s_0)\|}\mathbf{t}\cdot(\frac{\Delta s^2}{2!}\mathbf{r}''(s_0) + \cdots)$ is an error term due to the higher orders of Taylor series approximation.

If a point on a curve $\mathbf{r}(s_0+\Delta s)$ is indeed a closest point to the $\mathbf{p_e}$, then vectors $\mathbf{t}$ and $\mathbf{e}$ become perpendicular and the term $\mathbf{t}\cdot\mathbf{e}$ becomes zero. Henceforth, the error term $\Delta s_{e,p}$ becomes zero. In practice vectors $\mathbf{e}$ and $\mathbf{t}$ will not be entirely perpendicular, however it can be assumed that the error term $\Delta s_{e,p}$ will have a negligible effect. The assumption is also made that the increment Δs is small and higher order terms of Δs are small and henceforth the $\Delta s_{e,d}$ can also be neglected.

The final equation for increment Δs is then

$$\Delta s = \frac{1}{\|\mathbf{r}'(s_0)\|}\mathbf{t}\cdot\Delta\mathbf{r} \tag{10.13}$$

with the error

$$\Delta s_e = \Delta s_{e,p} + \Delta s_{e,d}. \tag{10.14}$$

Increment Δs is therefore a projection of a vector $\Delta\mathbf{r}$ at a tangent of the curve, weighted by the factor $\frac{1}{\|\mathbf{r}'(s_0)\|}$. The problem might arise if a term $\|\mathbf{r}'(s_0)\|$ becomes zero. However, since it was assumed that the curve has a non-zero tangent on the interval where it is defined, the term $\|\mathbf{r}'(s_0)\|$ will never become zero. In an actual implementation a virtual fixture defined geometrically by a curve with a zero tangent will also pose other practical problems and it is therefore a good practice to avoid such curves.

Algorithm for estimation of the closest point on a parametric curve

---

Initialize
$s_0 = s_s, \Delta s_0 = 0$

for $i \in \{1, \dots \infty\}$
calculate increment Δ *s for i-th step*
$\mathbf{r}'_i = \mathbf{r}'(s_{i-1})$
$\mathbf{t}_i = \frac{\mathbf{r}'_i}{\|\mathbf{r}'_i\|}$
$\Delta \mathbf{r}_i = \mathbf{p}_{e,i} - \mathbf{r}_i$
$\Delta s_i = \frac{1}{\|\mathbf{r}'_i\|} \mathbf{t} \cdot \Delta \mathbf{r}_i$
$s_i = s_{i-1} + \Delta s_i$
if $s_i > s_e$, *then* $s_i = s_e$
if $s_i < s_s$, *then* $s_i = s_s$
calculate error term Δs_e *for increment* Δs_i
$\mathbf{e}_i = \mathbf{p}_{e,i} - \mathbf{r}(s_i)$
$\Delta s_e = \frac{1}{\|\mathbf{r}'_i\|} \mathbf{t} \cdot \mathbf{e}_i + \frac{1}{\|\mathbf{r}'_i\|} \mathbf{t} \cdot (\frac{\Delta s_i^2}{2} \mathbf{r}''(s_{i-1}))$
if Δs_e *is too large, reiterate step i with new value of* s_i *instead of* s_{i-1}
$s_{i-1} = s_i$
else continue with step (i+1)

outputs of the algorithm in step i are
1) *closest point on curve* $\mathbf{r}(s_i)$ *of the end-effector position* $\mathbf{p}_{e,i}$
2) *tangent on the curve* $\mathbf{t}_i$

10.2.2 Virtual Fixtures Based Control

A controller transforms the user's input either into the movement of the robotic manipulator or into the end-effector force applied by the robotic manipulator. In case of an admittance controlled virtual fixture, the user's input is measured force and the output is desired movement of the robotic manipulator. In case of the impedance controlled virtual fixture, the input is measured position or velocity and the output of the virtual fixture is force of interaction between the end-effector with the virtual fixture.

The most appropriate selection for an admittance controlled robotic system is to use the admittance type virtual fixtures and for an impedance controlled robotic system is to use the impedance type virtual fixtures. However, an impedance controlled robotic system as well as an admittance type virtual fixture, can be used by employing pseudo-admittance control. Pseudo-admittance control will be discussed in Sect. 10.4.

10.2.2.1 Impedance Control

Impedance type guidance virtual fixtures are commonly implemented as force fields [10], which guide the movement toward a desired point or direction. The guidance virtual fixture, which guides the movement toward the desired point, is implemented as a point-attraction virtual fixture. The point-attraction virtual fixture pulls the end-effector toward the desired point. This is typically achieved with a spring-damper model

$$\mathbf{F}_{vf,attraction} = K\mathbf{d} - B\dot{\mathbf{p}}_e \tag{10.15}$$

$$\mathbf{d} = \mathbf{p}_d - \mathbf{p}_e \tag{10.16}$$

where $\mathbf{d}$ is a distance vector between the end-effector $\mathbf{p}_e$ and desired point $\mathbf{p}_d$, $\dot{\mathbf{p}}_e$ is velocity of the end-effector, K is stiffness and B is damping. The desired point can be the closest point on a curve or a final point to which the end-effector should move. If the distance between the end-effector and desired point is large, then attraction force $\mathbf{F}_{vf,attraction}$ can be exceedingly large. In that case it should be limited to a maximal allowed attraction force. Another approach is to interpolate movement from an initial end-effector position to a final desired position so that the desired point of attraction moves along the interpolated path. To limit amplitude of the attraction force the length of $\mathbf{d}$ is limited to a predefined maximal value of d_{max}. Furthermore, stiffness K can be made time-dependent to allow gradual temporal transition of the otherwise distance-dependent attraction force [11]

$$\mathbf{D}(\mathbf{d}) = \begin{cases} \mathbf{d} & \text{if } \|\mathbf{d}\| < d_{max} \\ \frac{\mathbf{d}}{\|\mathbf{d}\|} d_{max} & \text{if } \|\mathbf{d}\| \geq d_{max} \end{cases} \tag{10.17}$$

$$K(t) = \begin{cases} K\frac{t}{\tau} & \text{if } t < \tau \\ K & \text{if } t \geq \tau \end{cases} \tag{10.18}$$

$$\mathbf{F}_{vf,attraction} = K(t)\mathbf{D}(\mathbf{d}) - B\dot{\mathbf{p}}_e \tag{10.19}$$

The guidance virtual fixture, which guides the movement in a desired direction along the path or surface, generates a guidance force in the direction of a tangent to the curve or surface. The guidance force along the path is both time and distance dependent. When the guidance virtual fixture is initiated the time-component allows gradual temporal transition from zero guidance force to maximal allowed guidance force, while the distance-dependent component allows a gradual transition to zero value of the guidance force when the end-point of the manipulator approaches the final desired point

$$\mathbf{F}_{vf,tangent} = f_d f_t F_{max} \mathbf{u}, \tag{10.20}$$

where $\mathbf{u}$ is a unit tangent vector in the closest point on a curve that defines the path of the virtual fixture, F_{max} is a maximal guidance force along the path of the virtual fixture, f_d and f_t are distance-dependent and time-dependent functions

$$f_d(\mathbf{d}) = \begin{cases} 1 & \text{if } \|\mathbf{d}\| \geq d_{min} \\ \frac{\|\mathbf{d}\|}{d_{min}} & \text{if } \|\mathbf{d}\| < d_{min} \end{cases} \quad (10.21)$$

$$f_t(t) = \begin{cases} \frac{t}{\tau} & \text{if } t < \tau \\ 1 & \text{if } t \geq \tau \end{cases} \quad (10.22)$$

The force of the impedance guidance virtual fixture that guides the end-point of the robotic system along the desired path and also to the curve or surface is the sum of attraction force $\mathbf{F}_{vf,attraction}$ and guidance tangent force $\mathbf{F}_{vf,tangent}$

$$\mathbf{F}_{vf} = \mathbf{F}_{vf,attraction} + \mathbf{F}_{vf,tangent}. \quad (10.23)$$

10.2.2.2 Admittance Control

Virtual fixtures are especially suitable for admittance type robotic systems. Admittance controlled robotic systems are inherently stiff and can strictly forbid the motion inside the forbidden region or can constrain the motion to the desired path. In admittance control law the input is measured force applied by the user and the output is desired velocity of the robotic manipulator. Input force $\mathbf{F}$ is decomposed into the force applied in the preferred direction $\mathbf{F}_D$ and into the force applied in the non-preferred direction $\mathbf{F}_\tau$, which is orthogonal to the preferred direction. Output force $\mathbf{F}_{vf}$ from the virtual fixture is

$$\mathbf{F}_{vf} = \mathbf{F}_D + k_\tau \mathbf{F}_\tau. \quad (10.24)$$

The admittance control law for guidance virtual fixture can be written as

$$\mathbf{v} = \alpha(\mathbf{F}_D + k_\tau \mathbf{F}_\tau), \quad (10.25)$$

where α is an admittance that maps force into velocity and k_τ is admittance ratio. Admittance α is inversely proportional to the damping of the virtual fixture $b = \frac{1}{\alpha}$ and controls the overall compliance of the robotic manipulator to the input force. Admittance ratio $k_\tau \in [0, 1]$ controls the compliance of the virtual fixture in the non-preferred direction. If admittance ratio is set to zero $k_\tau = 0$, virtual fixture completely constrains the motion to the preferred direction and the virtual fixture is referred to as a *hard virtual fixture*. If admittance ratio is set to one $k_\tau = 1$, virtual fixture has no effect on the movement and compliance of the movement is equal in all directions. For cases where admittance ratio is set to values between 0 and 1, the virtual fixture is referred to as *soft virtual fixture*, since it is possible to move the manipulator in the non-preferred direction in an amount controlled by the admittance ratio k_τ. Note, that it is also possible to set the value of the admittance ratio k_τ to values above 1. In such a case the manipulator is more compliant in a non-preferred direction and a virtual fixture loses its function as an aid for guiding the movement along the preferred direction.

Admittance type virtual fixtures can be either passive or active.

Passive Virtual Fixture

In passive guidance virtual fixtures, only the motion along the preferred direction is maintained. In contrast to active virtual fixtures the path of movement guided by the passive guidance virtual fixtures can slowly deviate from the desired path. Although the passive virtual fixtures do not guide the movement back to the path or surface, in some cases this might be desirable. An example might be when the user needs to retain the control, to move the robotic system anywhere in the safe region of the workspace, but at the same time the controller should nevertheless retain the ability to attenuate movements in non-preferred directions by employing the passive guidance virtual fixtures and retaining the ability to guide the user.

First the force applied by the user in preferred direction $\mathbf{F}_D$ is calculated by projecting input force $\mathbf{F}$ to the tangent $\mathbf{t}$ (Fig. 10.4)

$$F_D = \mathbf{t} \cdot \mathbf{F} \tag{10.26}$$

$$\mathbf{F}_D = \mathbf{t} F_D. \tag{10.27}$$

Force in non-preferred direction $\mathbf{F}_\tau$ is calculated by subtracting the force in preferred direction $\mathbf{F}_D$ from force $\mathbf{F}$ applied by the user

$$\mathbf{F}_\tau = \mathbf{F} - \mathbf{F}_D. \tag{10.28}$$

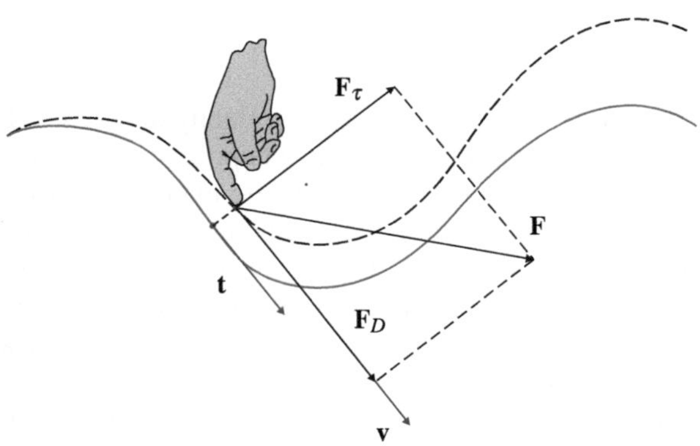

Fig. 10.4 Passive guidance virtual fixture

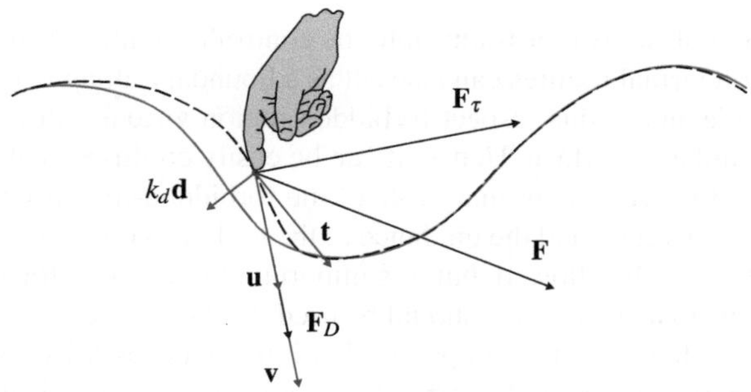

Fig. 10.5 Active virtual fixture

Active Virtual Fixture

When the end-effector deviates from the virtual fixture, active virtual fixture compensates for this deviation to guide the movement back to the path of the virtual fixture by adjusting the preferred direction in proportion to the amount of the deviation from the path.

First a vector $\mathbf{d}$ of the deviation from the path is calculated as a vector originating from the end-point position $\mathbf{p}_e$ and pointing toward the closest point on a curve $\mathbf{r}(s)$. Then a preferred direction designated with $\mathbf{u}$ is adjusted accordingly (Fig. 10.5)

$$\mathbf{d} = \mathbf{r}(s) - \mathbf{p}_e \tag{10.29}$$

$$\mathbf{U} = \mathbf{t} + k_d\mathbf{d} \tag{10.30}$$

$$\mathbf{u} = \frac{\mathbf{U}}{\|\mathbf{U}\|}, \tag{10.31}$$

where k_d is a stiffness factor of the active virtual fixture. The force applied in the preferred direction $\mathbf{F}_D$ and the force applied in the non-preferred direction $\mathbf{F}_\tau$ can be computed as

$$F_D = \mathbf{u} \cdot \mathbf{F} \tag{10.32}$$

$$\mathbf{F}_D = \mathbf{u}F_D \tag{10.33}$$

$$\mathbf{F}_\tau = \mathbf{F} - \mathbf{F}_D. \tag{10.34}$$

10.3 Forbidden-Region Virtual Fixtures

Forbidden-region virtual fixtures are used to constrain movements of the tool tip to the region in which the tool tip is allowed to move and to prevent those movements into the forbidden region. The main purpose of the forbidden-region virtual fixtures is implementation of safety constraints into the controller of the robotic system [12]. Forbidden-region virtual fixtures can also act as a boundary along which the tool tip is allowed to slide and in this respect forbidden-region virtual fixtures can also act as a form of a guidance surface. Hence, it can be easily confused with the guidance virtual fixtures. However, the primary role of the forbidden-region virtual fixture is to act as a safety constraint and the guidance of the tool tip is only a secondary effect. In some cases this can be allowed, but it is important to note that, for the purpose of guidance, guidance virtual fixtures should be used. Forbidden-region virtual fixtures can provide haptic feedback to the operator but it is not necessarily so.

A typical implementation of the forbidden-region virtual fixture is its use in surgical robotics in laparoscopic surgery. The tool tip typically maneuvers in a very tight region surrounded with a delicate tissue of the vital organs, which can be damaged if it comes into contact with the tool tip. Guidance virtual fixtures guide the tool tip to maneuver along the optimal path as farthest from the surrounding tissue as possible. However, in case of unexpected deviation the forbidden-region virtual fixtures constrain and prevent the tool tip from diverging into the forbidden region. Forbidden-region virtual fixture are therefore used in applications where the workspace of the manipulator must be restricted for safety reasons. Forbidden-region virtual fixtures can be implemented on both cooperative robotic systems and teleoperation robotic systems.

- Cooperative robotic systems. Since cooperative robotic systems are built with only a slave manipulator the forbidden-region virtual fixtures are implemented on the slave robotic manipulator.
- Teleoperation robotic system. Just as with the guidance virtual fixtures, forbidden-region virtual fixtures in a teleoperation robotic system can be implemented on the slave side, on the master side or on both sides. The aim of any type of implementation of forbidden-region virtual fixtures is to constrain the motion of the slave device to prevent penetration into the forbidden region on the slave side. Penetration of the master device into the forbidden region is of less importance, since the slave device is the device which is actually in contact with the environment and can therefore cause potentially unsafe manipulation. Therefore, the most obvious implementation of the forbidden-region virtual fixtures is on the slave side. However, implementations of forbidden-region virtual fixtures on the slave side, on the master side or on both sides can be found in teleoperation robotic systems [12, 13]. Again, for all three types of implementations the aim is the same: to prevent movement of the slave device into the forbidden region. The implementations differ in the case of haptic feedback to the user and level of disturbance rejection. The disturbance on the slave side will be best suppressed by the forbidden-region virtual fixtures on the slave side. The disturbance on the master side, however, will be equally well suppressed by the forbidden-region virtual fixtures on either master side or slave side. If the forbidden-region virtual fixtures are to be implemented on just one side, the best disturbance suppression will be achieved if a forbidden-region virtual fixture is implemented on the slave side. Implementation of the forbidden-region virtual fixtures on both sides will

result in better suppression of the disturbance originating on either side. In case of implementing forbidden-region virtual fixtures on both sides, the forbidden-region virtual fixtures implemented on the slave side will make a larger contribution to the disturbance suppression than forbidden-region virtual fixtures implemented on the master side. However, the addition of the forbidden-region virtual fixture on the master side will result in an improved experience of a telepresence due to the increased perceived stiffness of the virtual fixture and smaller error position between master and slave device [13].

Again, as with the guidance virtual fixtures, also the forbidden-region virtual fixtures can be of an impedance or admittance type.

Impedance type forbidden-region virtual fixtures

The simplest and most common implementation of the forbidden-region virtual fixtures is a virtual wall. A virtual wall is an impedance surface and the force needed to penetrate into the virtual wall is proportional to the penetrated depth. A unilateral high gain PD type controller or a spring-damper surface is commonly used to oppose the movement into the forbidden region (Sect. 6.2) when the tool tip is inside the restricted region, but they have no effect outside the restricted region. This is a penalty based method and requires a penetration of the end-effector into the forbidden region to activate the forbidden-region virtual fixtures.

Admittance type forbidden-region virtual fixtures

The admittance control law is

$$\mathbf{v} = \alpha(\mathbf{F}_D + k_\tau \mathbf{F}_\tau), \tag{10.35}$$

where α is an admittance, which maps force into velocity and k_τ is the admittance ratio. In admittance type forbidden-region virtual fixtures the non-preferred direction $\mathbf{F}_\tau$ is a direction in an opposite direction of the surface normal of the forbidden-region virtual fixtures. In the guidance virtual fixtures case, first the preferred direction $\mathbf{F}_D$ is calculated and then the non-preferred direction $\mathbf{F}_\tau$ is determined, perpendicular to the preferred direction $\mathbf{F}_D$. However, in the forbidden-region virtual fixtures case the opposite is done. When the tool tip moves to the surface of the forbidden region, the forbidden-region virtual fixtures are engaged. First a non-preferred direction $\mathbf{F}_\tau$ is calculated from the normal $\mathbf{n}$ to the surface. The normal $\mathbf{n}$ to the surface is a unit vector on the surface pointing from the forbidden region into the allowed region and is perpendicular to the surface. A non-preferred direction $\mathbf{F}_\tau$ is therefore a vector pointing inside the forbidden region in an opposite direction of the normal vector $\mathbf{n}$

$$F_\tau = \begin{cases} \mathbf{F}\cdot\mathbf{n} & \text{if } \mathbf{F}\cdot\mathbf{n} < 0 \\ 0 & \text{if } \mathbf{F}\cdot\mathbf{n} \geq 0 \end{cases}, \tag{10.36}$$

$$\mathbf{F}_\tau = \mathbf{n} F_\tau. \tag{10.37}$$

Then the preferred direction is calculated. The preferred direction is not simply a direction perpendicular to the non-preferred direction, but depends on the direction of the force applied by the user. If the force points inside the forbidden region (case where $\mathbf{F} \cdot \mathbf{n} < 0$), then the preferred direction is indeed perpendicular to the non-preferred direction. However, if the force points in a direction outside the forbidden region (the case where $\mathbf{F} \cdot \mathbf{n} \geq 0$), defining the preferred direction perpendicular to the non-preferred direction would not allow the tool tip to move off the surface into the safe region. In that case a tool tip should be allowed to move from the surface. The preferred direction is therefore in a direction of the applied force and the tool tip is free to move outside the forbidden region

$$\mathbf{F}_D = \mathbf{F} - \mathbf{F}_\tau. \tag{10.38}$$

Since the fundamental property of the forbidden-region virtual fixtures is to fully forbid penetration inside the restricted region, the forbidden-region virtual fixtures is implemented as a hard virtual fixture: the admittance ratio k_τ is set to zero and output force from the virtual fixture becomes $\mathbf{F}_{vf} = \mathbf{F}_D$. The control law is then

$$\mathbf{v} = \alpha \mathbf{F}_D \tag{10.39}$$

The control law eliminates the component of movement directed inside the forbidden-region virtual fixtures.

10.4 Pseudo-Admittance Bilateral Teleoperation

Pseudo-admittance bilateral teleoperation is based on use of a virtual agent called a proxy. The slave device follows the proxy, which has an admittance type dynamics, while the master device follows the slave device. The force of the environment measured at the slave device is scaled and conveyed to the master device [14].

Since the proxy is a software-generated virtual agent, the dynamics of the proxy can be arbitrarily chosen based on the task requirements. In an unconstrained environment, proxy movement depends on the error between the slave and the master device. However, in a constrained environment, proxy movements can be directed by the virtual fixtures. Figure 10.6 shows the key difference between the pseudo-admittance control law, which uses a proxy, and simple teleoperation, which uses master-slave virtual coupling.

When a slave moves in a free space, the velocity of the slave is approximately proportional to the force applied by the user on the master device and therefore pseudo-admittance control imitates the admittance control. The pseudo-admittance control has the property of a quasi-static transparency, and the force reflected back to the user is at low velocity of the system, approximately the same as the force between the slave and the environment. Therefore at zero velocity there is a complete

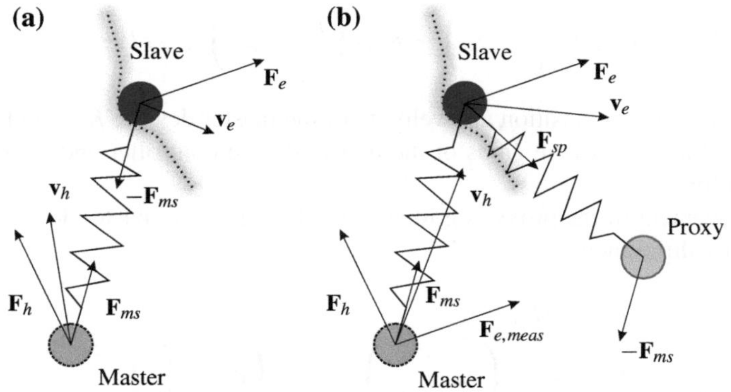

Fig. 10.6 **a** Simple bimanual teleoperation with virtual coupling between master and slave. The sum of interaction force $\mathbf{F}_e$ between the environment and end-point of the slave and master-slave virtual coupling force $-\mathbf{F}_{ms}$ governs the dynamics of the slave, while force applied by the user $\mathbf{F}_h$ and master-slave virtual coupling force $\mathbf{F}_{ms}$ governs the dynamics of the master. The user feels the compliance of the environment through the master-slave coupling. **b** The bilateral teleoperation employing the pseudo-admittance control law. The measured environment force $\mathbf{F}_{e,meas}$ is conveyed directly to the user. The master follows the slave, while the slave follows the proxy. Proxy dynamics is governed by the force of the master-slave virtual coupling

position correspondence between the slave and master positions and an interaction force between the slave and the environment is perfectly reflected to the user [14].

10.4.1 Impedance Type Master and Impedance Type Slave

The pseudo-admittance control law can be used in a bilateral manipulation where both the master and the slave devices are of impedance type. The same law imitates the admittance control system on both slave and master impedance type devices.

The proportional control law with velocity feedback for the slave device ensures that the slave device follows the proxy

$$F_{cs} = K_{ps}(\gamma_v p_p - p_s) - K_{ds}\dot{p}_s, \tag{10.40}$$

where p_p is the position of the proxy, p_s and $\dot{p}_s$ are position and velocity of the slave device, K_{ps} and K_{ds} are proportional and derivative gains of the slave device controller and γ_v is a velocity scaling factor.

The proportional-derivative control law with feedforward measured environment interaction force ensures that the master device follows the slave device and that the environment force is reflected to the user as

$$F_{cm} = K_{pm}\left(\frac{p_s}{\gamma_v} - p_m\right) + K_{dm}\left(\frac{\dot{p}_s}{\gamma_v} - \dot{p}_m\right) + \gamma_f F_e, \tag{10.41}$$

where p_m and $\dot{p}_m$ are position and velocity of the master device, K_{pm} and K_{dm} are proportional and derivative gains of the master device controller and γ_f is a force scaling factor.

The movement of the proxy is governed by the slave and master device position and velocity difference

$$\dot{p}_p = \alpha F_p \tag{10.42}$$

$$F_p = K_{pm}\left(p_m - \frac{p_s}{\gamma_v}\right) + K_{dm}\left(\dot{p}_m - \frac{\dot{p}_s}{\gamma_v}\right), \tag{10.43}$$

where α is admittance, which determines the proxy dynamics.

Rewriting Eqs. (10.41) and (10.40) yields the parameters of subsystems of the block diagram shown on Fig. 10.7a

$$F_{cm} = \frac{1}{\gamma_v}\frac{K_{dm}s + K_{pm}}{s}\dot{p}_s - \frac{K_{dm}s + K_{pm}}{s}\dot{p}_m + \gamma_f F_e$$

$$F_{cs} = K_{ps}\gamma_v\frac{\alpha}{s}F_p - \frac{K_{ds}s + K_{ps}}{s}\dot{p}_s$$

$$\begin{aligned} C_m &= \frac{K_{dm}s + K_{pm}}{s} \\ C_s &= \frac{K_{ds}s + K_{ps}}{s} \\ C_2 &= \gamma_f \\ C_3 &= K_{ps}\gamma_v\frac{\alpha}{s} \\ C_4 &= \frac{1}{\gamma_v}\frac{K_{dm}s + K_{pm}}{s} \end{aligned} \tag{10.44}$$

The block diagram shown on Fig. 10.7a is a modified version of the general four-channel bilateral teleoperation control scheme described in Chap. 9. The scheme is modified to allow the use of a pseudo-admittance control scheme.

The slave is connected to the proxy through the slave-proxy virtual coupling and the master is connected to the slave through the master-slave virtual coupling. Slave device dynamics is governed by the slave-proxy virtual coupling force and environment force and by the velocity feedback damping. The measured interaction force between the environment and slave is forwarded to the master. The master device dynamics are governed by the sum of the master-slave virtual coupling force, measured environment force and force applied by the user. The dynamics of the proxy is governed by the force of the master-slave virtual coupling.

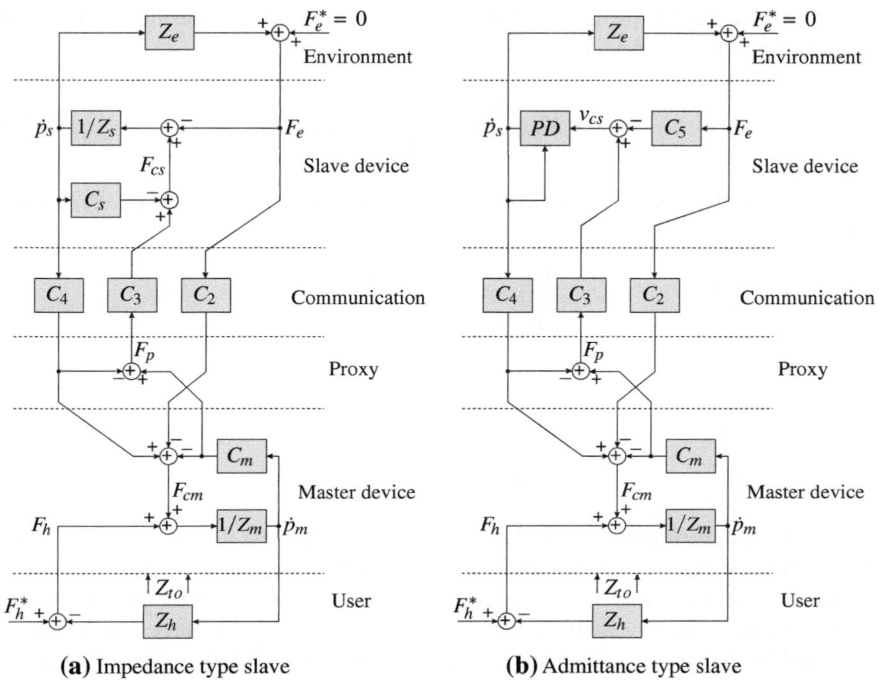

Fig. 10.7 The figure shows a teleoperation system with the pseudo-admittance bilateral control law. **a** A teleoperation with impedance type master and slave device, while **b** a system with impedance type master device and slave admittance type device

10.4.2 Impedance Type Master and Admittance Type Slave

In an admittance type slave the low-level PD controller assures that the slave device closely follows the commanded velocity v_{cs},

$$v_{cs} = \alpha(F_p - F_e) \tag{10.45}$$

The proportional control law for a master device with velocity feedback and feedforward measured environment force ensures that the master device follows the slave device

$$F_{cm} = K_{pm}\left(\frac{p_s}{\gamma_v} - p_m\right) - K_{dm}\dot{p}_m + \gamma_f F_e \tag{10.46}$$

$$F_{cm} = \frac{1}{\gamma_v}\frac{K_{pm}}{s}\dot{p}_s - \frac{K_{dm}s + K_{pm}}{s}\dot{p}_m + \gamma_f F_e. \tag{10.47}$$

The movement of the proxy is a function of the slave and master device position difference and master device velocity

$$F_p = K_{pm}\left(p_m - \frac{p_s}{\gamma_v}\right) + K_{dm}\dot{p}_m. \tag{10.48}$$

Parameters of the subsystem blocks of the block diagram shown on Fig. 10.7b are

$$\begin{aligned} C_m &= \frac{K_{dm}s + K_{pm}}{s} \\ C_2 &= \gamma_f \\ C_3 &= \alpha \\ C_4 &= \frac{1}{\gamma_v}\frac{K_{pm}}{s} \\ C_5 &= \alpha. \end{aligned} \tag{10.49}$$

The desired movement of the slave device is governed by the measured environment force and the force of the master-slave virtual coupling. The measured interaction force between the environment and slave is forwarded to the master. Master device dynamics is governed by master-slave virtual coupling force, environment force and by the velocity feedback damping. The dynamics of the proxy is governed by the force of the master-slave virtual coupling and damping proportional to master device velocity.

10.4.3 Virtual Fixtures with Pseudo-Admittance Control

In pseudo-admittance control, virtual fixtures are applied on the proxy and not on the slave device. Virtual fixtures thus directly act on the proxy. The input to the virtual fixtures is the force F_p of the master-slave virtual coupling and the output from the virtual fixture is force $F_{p,vf}$, which governs the proxy dynamics.

In teleoperation system with impedance type master and slave device (Fig. 10.7a) Eq. (10.42) becomes

$$\dot{p}_p = \alpha F_{p,vf}. \tag{10.50}$$

In a teleoperation system with impedance type master and admittance type slave devices (Fig. 10.7b) Eq. (10.45) becomes

$$v_{cs} = \alpha(F_{p,vf} - F_e). \tag{10.51}$$

Figure 10.8 shows an example of a teleoperation system with impedance type master and admittance type slave device.

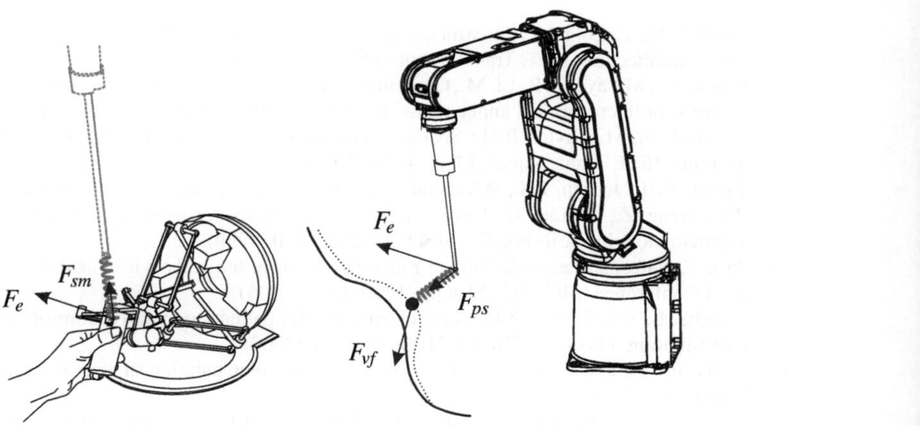

Fig. 10.8 The figure shows the bilateral teleoperation employing the pseudo-admittance control law with impedance type master and admittance type slave. Environment force F_e is conveyed directly to the user. The master follows the slave, while the slave follows the proxy. Proxy dynamics is governed by the virtual fixture force F_{vf}

Under a pseudo-admittance control scheme, the velocity of both the master and slave device are approximately proportional to the force applied by the user on the master and force of the environment on the slave. Pseudo-admittance control also displays quasi-static transparency, which approaches perfect transparency, when velocity approaches zero. Virtual coupling between the slave and proxy and virtual coupling between the slave and the master act as low-pass filters, which suppress high frequency components of the user's movements such as tremor and other unwanted and uncontrolled faster movements. On the other hand, by directly linking the force between the slave device and environment to the user, the user has access to the high-bandwidth haptic information of the interaction between the slave device and the environment [14].

References

1. Abbott, J.J., Marayong, P., Okamura, A.M.: Haptic virtual fixtures for robot-assisted manipulation. In: 12th International Symposium of Robotics Research (ISRR), pp. 49–64 (2005)
2. Rosenberg, L.B.: Virtual fixtures: perceptual tools for telerobotic manipulation. In: Virtual Reality Annual International Symposium, pp. 76–82 (1993)
3. Payandeh, S., Stanisic, Z.: On application of virtual fixtures as an aid for telemanipulation and training. In: Proceedings of the 10th Symposium on Haptic Interfaces for Virtual Environment and Teleoperator Systems (HAPTICS.02), pp. 18–23 (2002)
4. Hager, G.: Human-machine cooperative manipulation with vision-based motion constraints. In: Chesi, G., Hashimoto, K. (eds.) Visual Servoing via Advanced Numerical Methods. Lecture Notes in Control and Information Sciences, vol. 401, pp. 55–70. Springer, Berlin (2010)

5. Peshkin, M.A., Colgate, J.E., Wannasuphoprasit, W., Moore, C.A., Gillespie, R.B., Akella, P.: Cobot architecture. IEEE Trans. Robot. Autom. **17**(4), 377–390 (2001)
6. Kragic, D., Marayong, P., Li, M., Okamura, A.M., Hager, G.D.: Human-machine collaborative systems for microsurgical applications. Int. J. Robot. Res. **24**(9), 731–741 (2005)
7. Li, M., Ishii, M., Taylor, R.H.: Spatial motion constraints using virtual fixtures generated by anatomy. IEEE Trans. Robot. **23**(1), 4–19 (2007)
8. Taylor, R.H., Jensen, P.S., Whitcomb, L., Barnes, A.C., Kumar, R., Stoianovici, D., Gupta, P.K., Wang, Z., de Juan, E., Kavoussi, L.R.: A steady-hand robotic system for microsurgical augmentation. Int. J. Robot. Res. **18**(12), 1201–1210 (1999)
9. Park, S.: Safety strategies for human-robot interaction in surgical environment. In: International Joint Conference (SICE-ICASE), pp. 1769–1773 (2006)
10. Mihelj, M., Nef, T., Riener, R.: A novel paradigm for patient-cooperative control of upper-limb rehabilitation robots. Adv Robot **21**(8), 843–867 (2007)
11. Prada, R., Payandeh, S.: On study of design and implementation of virtual fixtures. Virtual Reality **13**, 117–129 (2009)
12. Abbott, J.J., Okamura, A.M.: Stable forbidden-region virtual fixtures for bilateral telemanipulation. J. Dyn. Syst. Meas. Control **126**(1), 53–64 (2006)
13. Abbott, J.J., Okamura, A.M.: Virtual fixture architectures for telemanipulation. In: Proceedings of IEEE International Conference on Robotics and Automation (ICRA '03), vol. 2, pp. 2798–2805 (2003)
14. Abbott, J.J., Okamura, A.M.: Pseudo-admittance bilateral telemanipulation with guidance virtual fixtures. Intl. J. Robot. Res. **26**(8), 865–884 (2007)

Chapter 11
Micro/Nanomanipulation

Handling of and interaction with increasingly smaller objects, where human sensing and manipulation capabilities are no longer sufficient, is becoming an important area of robotics. Due to fast development in the area of nanotechnologies, handling of materials on the molecular level is becoming imperative. With the precise positioning of atoms, molecules and objects of nanoscale dimensions it is possible to construct new sensors, materials with specific characteristics, robots and devices of micrometer dimensions. However, development of nanotechnology requires previous solution of some basic problems. Nanomanipulation is one of the most important. Physical and chemical phenomena at the level of nanoscale are still not well understood. Therefore, new tools, control algorithms, sensing technologies and user interfaces specific for the nanoworld need to be designed.

Nanomanipulation can be defined as manipulation of objects of nanoscale dimensions with tools of nanoscale dimensions and with sub-nanometer precision. Manipulation involves pushing, pulling, cutting, pick-and-place manipulations, orienting, assembly, bending and grooving of objects using a sensor-based feedback loop [1]. Some basic manipulations are shown in Fig. 11.1.

Nanomanipulation is becoming relevant for different scientific disciplines:

- *biotechnology*: nanomanipulation allows local and accurate manipulation of biological objects such as DNA or proteins; in addition to manipulation it also enables precise measurements of static and dynamic characteristics of biological structures,
- *micro/nanotechnology*: nanomanipulation enables assembly of nanoscale objects for construction of complex devices; with precise positioning of particles, nanotubes and molecules it is possible to construct electronic and quantum optical devices,
- *material science*: nanomanipulation allows construction of new materials and analysis of material properties such as friction, adhesion or electric properties at nanoscale.

M. Mihelj and J. Podobnik, *Haptics for Virtual Reality and Teleoperation*, Intelligent Systems, Control and Automation: Science and Engineering 64,
DOI: 10.1007/978-94-007-5718-9_11,

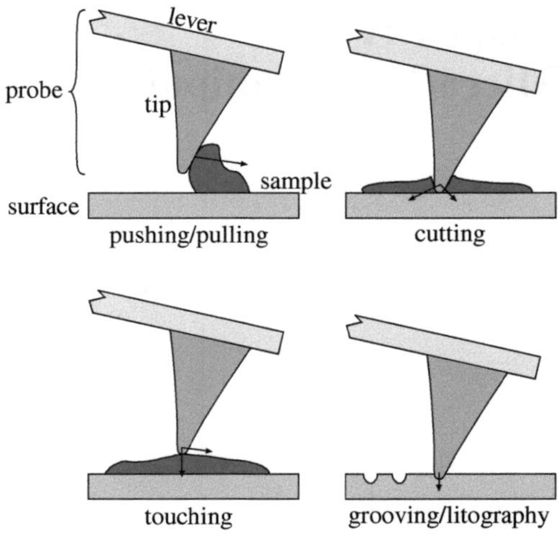

Fig. 11.1 Some mechanical micromanipulations [2]

Table 11.1 Scaling factors for different physical quantities (L indicates object length)

Quantity	Scaling factor
Length	L
Surface	L^2
Volume	L^3
Mass	L^3
Gravitational force	L^3
Inertial force	L^4
Linear spring constant	L
Oscillating frequency	$L^{-\frac{3}{2}}$

11.1 Nanoscale Physics

During the transition from the macro world to the world of micro/nano dimensions, the most obvious phenomenon is the reduction of the size of objects, where the change of length results in various scaling effects. Scaling factors for different physical and geometrical properties are summarized in Table 11.1.

It can be noted that by reducing the size of objects, inertial forces are significantly reduced, while the system oscillating frequency increases with smaller size. By taking into account the scaling effect, we can analyze changes that occur at the nanoscale:

- Surface and adhesion forces, such as van der Waals, electrostatic and capillary forces prevail over inertial forces. As a result sticking occurs.
- At the nanoscale, quantum mechanics and chemistry need to be taken into account. Newtonian physics becomes less relevant.
- Oscillating frequency increases with the decreasing object length, meaning that object's dynamics becomes faster at nanoscale.

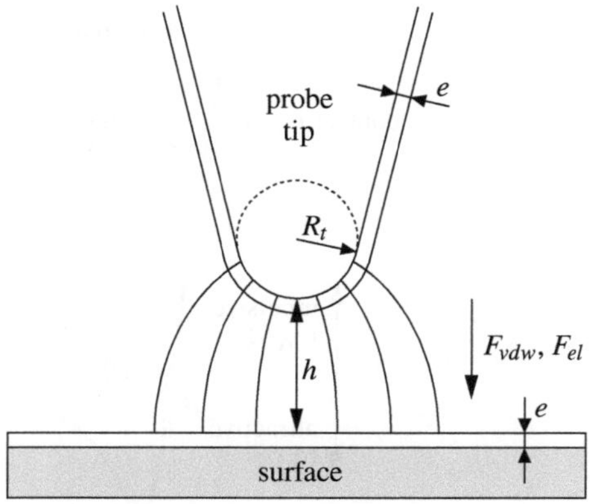

Fig. 11.2 Interaction between a probe tip and the sample via van der Waals, electrostatic and capillary forces [3]

- Adhesion forces depend on object geometry and material, on the distance between the probe and the object, on the environment parameters, such as temperature and humidity, and on the type of the environment (vacuum, air, liquid). Thus design of nanomanipulator, manipulation strategy, sensing and control highly depend on the specific task and the environment.
- Nanomanipulation control is sensitive to disturbances from the environment and uncertain environmental parameters may have unpredictable effects.

11.1.1 Model of Nanoscale Forces

With the assumption of a spherical (parabolic) shape of the nanomanipulation probe tip, we can assume forces between the sphere and the surface as shown in Fig. 11.2. The curve showing the relationship between force and distance of the probe from the sample is shown in Fig. 11.3.

During the probe approach to the sample a contactless attractive force appears at point A. The force that appears in a region without contact is the result of van der Waals and electrostatic long range forces. Due to the active force the probe tip finally touches the sample. Further approach is prevented due to Pauli's short range repulsive forces, which cause the probe cantilever to bend. The probe is in the contact region. While the probe retracts from the surface, a hysteresis can be observed due to the adhesion between the tip and the sample. The highest adhesion force appears after the transition through the point B. The probe tip completely separates from the surface at point C.

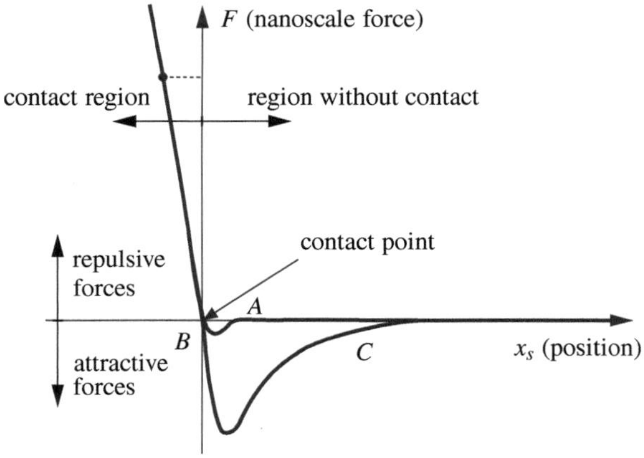

Fig. 11.3 A curve relating force and distance during approach and retraction from the surface

11.2 Nanomanipulation Systems

A typical nanomanipulation system shown in Fig. 11.4 consists of a nanomanipulator, sensors, a control subsystem and a user interface composed of haptic and visual displays.

11.2.1 Nanomanipulator

A nanomanipulator allows application of external forces on a sample. For manipulation of nanoscale objects a probe of similar dimensions is required. An *atomic force microscope* (AFM) and a *scanning tunneling microscope* (STM) are the most widespread types of nanomanipulators. Using the STM it is possible to manipulate atoms and molecules with the application of voltage pulses between the probe and the conducting sample. On the other hand, the AFM can be used for execution of mechanical tasks as shown in Fig. 11.1.

The principle of operation of an AFM microscope is shown in Fig. 11.5. Construction of an AFM is based on a probe with a flexible cantilever, a sensor for measuring cantilever bending with atomic resolution and a high precision Cartesian positioner [4]. The cantilever with the length of 100–400 μm and width of a few micrometers has a precisely calibrated spring constant k_c and resonant frequency f_r. The probe tip can be of conical, pyramidal or cylindrical shape with a diameter of a few 10 nm. Lever deformation resulting from interaction forces between the tip and the sample can be measured using a laser beam and photocells.

Due to attractive and repelling short or long range atomic forces F_z^c, the bending of the lever ζ depends on the distance between the probe tip and the sample L_z. If a probe is moved slowly enough, such that the lever is balanced after each displacement, the force applied on the sample F_z^c can be computed as a function of the lever deflection.

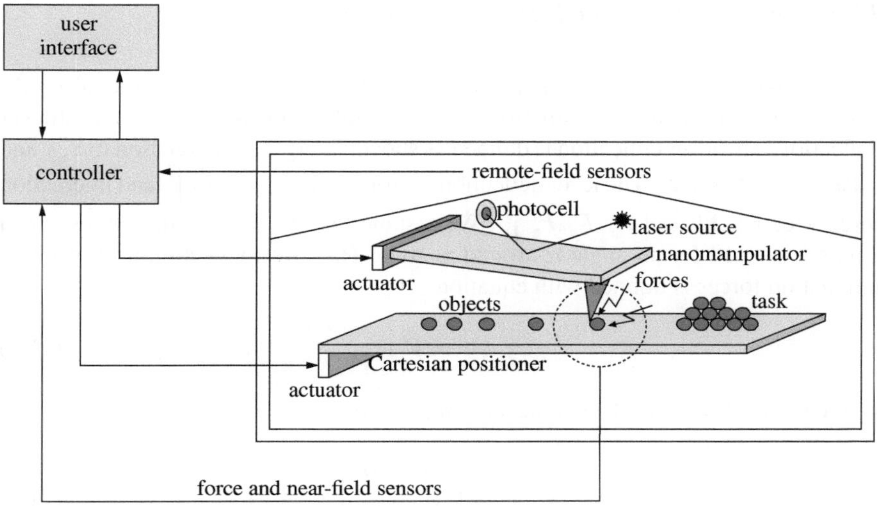

Fig. 11.4 Basic structure of a nanomanipulation system [1]

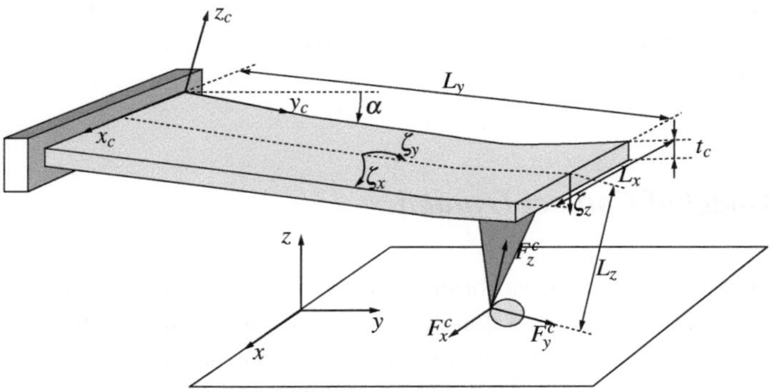

Fig. 11.5 Structure and operational principle of AFM microscope [2]

11.2.2 Actuators

Positioning and orienting of the sample and application of external forces on the sample requires actuation. Important characteristics of the actuators are their precision, range of motion, degrees of freedom and bandwidth. Drive precision is determined by its resolution, linearity, repeatability and mechanical and thermal noise. The most often used actuators for nanomanipulation are piezoelectric crystals and micro-electromechanical (electrostatic, thermic, capacitive) drives.

11.2.3 Measurement of Interaction Forces

Force sensing provides most of the feedback data during nanomanipulation. As already mentioned, force estimation can be based on measurement of cantilever deflection. The lever deflection is defined as shown in Fig. 11.5 in relation to x, y and z axes as ζ_x, ζ_y and ζ_z. The lever deflection vector is $\zeta = \left[\, \zeta_x \; \zeta_y \; \zeta_z \,\right]^T$ and interaction force vector is $\mathbf{F}^c = \left[\, F_x^c \; F_y^c \; F_z^c \,\right]^T$. We assume that the lever with stiffness k_c in z axis is rotated for an angle α around x_c axis. The lever deflection as a result of interaction forces is defined with equation

$$\zeta = \mathbf{C}\mathbf{F}^c, \tag{11.1}$$

where matrix $\mathbf{C}$ defines the cantilever compliance

$$\mathbf{C} = \frac{1}{k_c}\begin{bmatrix} c_1 & 0 & 0 \\ 0 & c_2 & c_3 \\ 0 & c_3 & 1 \end{bmatrix}. \tag{11.2}$$

Parameters c_1, c_2 and c_3 depend on the cantilever geometry. Relation (11.1) can be used for estimation of interaction forces based on measurements of cantilever deflections [2].

11.2.4 Model of Contact Dynamics

A model of nanoscale forces was briefly addressed in Sect. 11.1.1. A reliable control of nanomanipulation requires a detailed model of a nanomanipulator probe. When approaching or retracting the probe tip to/from a flat surface, the cantilever dynamics can be approximated using a simple mass, damper and spring system as shown in Fig. 11.6 [2]. P defines the actual load force, m_c, b_c and k_c represent lever mass, viscosity and spring constant, ζ_z defines the cantilever deflection in the vertical direction. Parameters $k_i(z, \zeta_z)$ and b_i are constants that define interaction stiffness and viscosity between the probe and the sample and δ defines the penetration depth. Lever dynamics can be written as

$$\begin{aligned} &m_c^*\ddot{\zeta}_z + 2m_c^*b_c\dot{\zeta}_z + k_c\zeta_z \\ &\quad = k_i(z, \zeta_z)(z - \zeta_z) + 2m_c^*b_i(\dot{z} + \dot{\zeta}_z), \end{aligned} \tag{11.3}$$

where $m_c^* = 0.24\, m_c$ is the effective mass of a lever with a rectangular cross section. With the assumption that the viscous effects are negligible, the dynamic model can be simplified to

$$m_c^*\ddot{\zeta}_z + k_c\zeta_z = k_i(z, \zeta_z)(z - \zeta_z)\,. \tag{11.4}$$

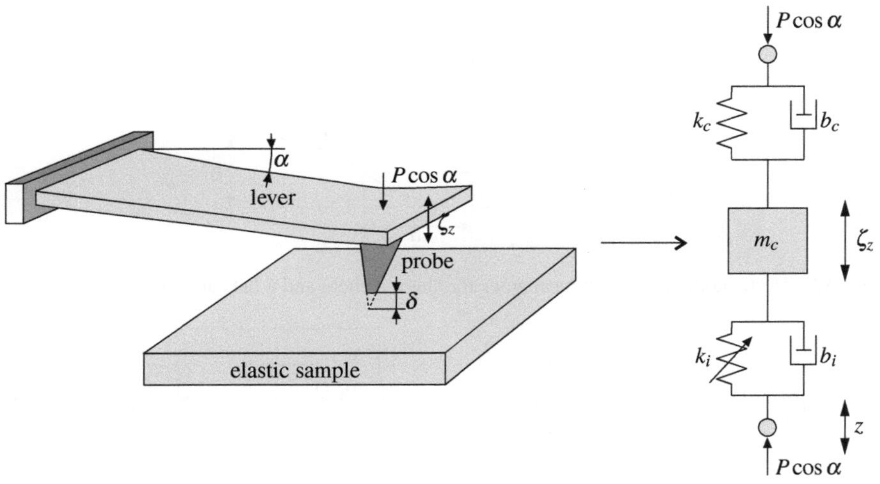

Fig. 11.6 A lever model consisting of a mass, a damper and a spring (adapted from [2])

Contact dynamics can be represented using various models. The most simple is the Hertz model of contact shown in Fig. 11.7. A detailed description of a contact model is beyond the scope of this chapter and will not be addressed here.

11.3 Control of Scaled Bilateral Teleoperation

Control of nanomanipulation can either be performed (1) via teleoperation or (2) completely automatically. In the case of teleoperation, the human operator is integrated into the control loop and performs nanomanipulation tasks via a user interface. A user interface is in this case composed of a visual display and a haptic interface. Teleoperation allows execution of complex tasks that require human cognitive capabilities and adaptability. Drawbacks of direct teleoperation are its slow execution, low accuracy and repeatability. In the case of automatic control the nanorobot is inserted into the control loop and control is based on sensor measurements without intervention of the operator. The drawback of the automatic control is poor reliability due to the complex dynamics of nanoscale objects, problems related to precise positioning, variable and uncertain physical parameters and inadequate models for implementation of control strategies.

A system shown in Fig. 11.8 allows scaled bilateral teleoperation, where nanoscale forces are mapped to the human arm [2]. The operator controls the nanomanipulator probe position via a haptic interface and at the same time perceives forces being applied by the probe on the sample.

Here we assume a haptic interface with a single degree of freedom. The operator manipulates the haptic display end-effector, of which pose z_m is measured. A probe

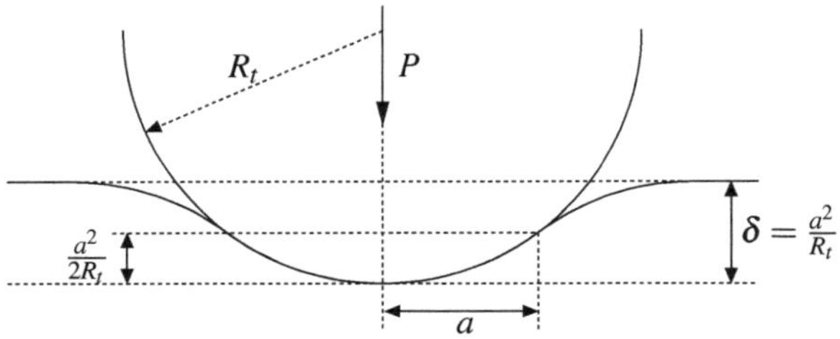

Fig. 11.7 Hertz model of contact mechanics for elastic sphere and a flat surface at load P

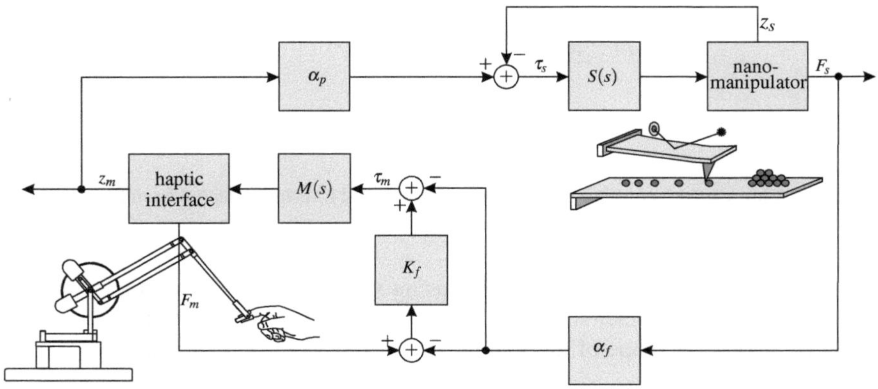

Fig. 11.8 Control of scaled bilateral teleoperation

vertical displacement z_s is controlled via a proportional-integral (PI) controller so that it tracks the scaled position of the haptic device $\alpha_p z_m$. Probe position z_s results in a nanoscale force F_s on the sample surface. The difference between the applied force on the haptic display and the scaled force $\alpha_f F_s$ is used in a proportional-derivative (PD) controller for computing the torque to be applied by the actuator of the haptic display, enabling haptic feedback.

11.3.1 Dynamic Model

The force being perceived by the user at the contact point with the haptic display is the result of the actuator torque transmitted to the human arm across the dynamics of the haptic device

$$m_m \ddot{z}_m + b_m \dot{z}_m = \tau_m + F_m, \tag{11.5}$$

where m_m is mass and b_m is the damping coefficient of the display, F_m and z_m are force and position of the operator and τ_m is the actuator torque.

On the nanomanipulator side are piezoelectric actuators and a laser-based position measuring system. Dynamics of the nanomanipulator z axis can be approximated with the following equation

$$\frac{1}{\omega_n^2}\ddot{z}_s + \frac{1}{\omega_n Q}\dot{z}_s + z_s = \tau_s - F_s, \tag{11.6}$$

where $\omega_n = 2\pi f_n$, f_n is the resonant frequency, Q is the quality factor, z_s determines the initial pose of the cantilever, F_s is the force being applied by the probe on the sample and τ_s is the force of the piezoelectric actuator.

Interaction dynamics defined by Eq. (11.4) can be written as

$$F_s = k_c\zeta_z = -m_c^*\ddot{\zeta}_z + k_i z_s - k_i\zeta_z, \tag{11.7}$$

where $z_s = z$. With the assumption of quasistatic (slow) motion, where after each displacement an equilibrium state z_s can be assumed, we can write

$$F_s = \frac{k_i(z_s, \delta)k_c}{k_i(z_s, \delta) + k_c} z_s \, . \tag{11.8}$$

11.3.2 Controller Design

An ideal controller response can be assumed as

$$\begin{aligned} z_s &\rightarrow \alpha_p z_m, \\ F_m &\rightarrow \alpha_f F_s \end{aligned} \tag{11.9}$$

in steady state. Constants $\alpha_f > 0$ and $\alpha_p > 0$ determine the scaling factors for position and force. With the assumption of a haptic force display, the controller equations can be written as [2]

$$\begin{aligned} \tau_m &= -\alpha_s F_s - K_f(\alpha_f F_s - F_m) \\ \tau_s &= K_d(\alpha_p \dot{z}_m - \dot{z}_s) + K_p(\alpha_p z_m - z_s), \end{aligned} \tag{11.10}$$

where K_p and K_d are controller proportional and derivative gains and K_f is the force error gain. Using the equations describing the haptic display and nanomanipulator dynamics and by assuming high stiffness values for interaction between the probe and the sample, $k_i \gg k_c$ and $F_s = k_c z_s$, we can determine relations in steady state as a result of an ideal response as

$$\frac{z_s}{z_m} = \alpha_p \frac{K_p}{1 + K_p},$$
$$\frac{F_m}{F_s} = \frac{1 + K_p}{k_c \alpha_p K_f K_p} + \alpha_f \left(1 + \frac{1}{K_f}\right). \qquad (11.11)$$

In order to achieve an ideal response it is necessary to select high gains K_p and K_f, while taking into account stability criteria.

References

1. Sitti, M.: Survey of nanomanipulation systems. In: Proceedings of the IEEE-Nanotechnology Conference, pp. 76–80 (2001)
2. Sitti, M., Hashimoto, H.: Teleoperated touch feedback from the surface at the nanoscale: modeling and experiments. IEEE/ASME Trans. Mechatron. **8**, 287–298 (2003)
3. Szemes, P.T., Ando, N., Korondi, P., Hashimoto, H.: Telemanipulation in the virtual nano reality. Trans. Autom. Control Comput. Sci. **45**, 117–122 (2000)
4. Sitti, M., Hashimoto, H.: Controlled pushing of nanoparticles: modeling and experiments. IEEE/ASME Trans. Mechatron. **5**, 199–211 (2000)

Index

M. Mihelj and J. Podobnik, *Haptics for Virtual Reality and Teleoperation*,
Intelligent Systems, Control and Automation: Science and Engineering 64,
DOI: 10.1007/978-94-007-5718-9, © Springer Science+Business Media Dordrecht 2012

MIX
Papier aus verantwortungsvollen Quellen
Paper from responsible sources
FSC® C105338

If you have any concerns about our products,
you can contact us on
ProductSafety@springernature.com

In case Publisher is established outside the EU,
the EU authorized representative is:
Springer Nature Customer Service Center GmbH
Europaplatz 3, 69115 Heidelberg, Germany

Printed by Libri Plureos GmbH
in Hamburg, Germany